The NEPA Dilemma

Outdated Environmental Rules and
the Surprising Ways They Cost You

Douglas B Sims, PhD

The NEPA Dilemma

Douglas B Sims, PhD

Copyright © 2024 by Douglas B. Sims

All rights reserved. No part of this book may be used or reproduced in any form whatsoever without written permission except in the case of brief quotations in critical articles or reviews.

Printed in the United States of America.

For more information, or to book an event, contact:
dsims@simsassociates.net

Book design: DB Sims
Cover design: DB Sims
Cover Photograph: Vladimir Timofeev through iStock

ISBN – Paperback: 979-8-9919108-8-0
ISBN – eBook: 979-8-9919108-9-7

First Edition: December 2024

The NEPA Dilemma

Douglas B Sims, PhD

Table of Contents

Acknowledgements... vii
Forward... ix
    1. The NEPA Dilemma in Telecommunications....................... 1
    2. The Origins and Goals of NEPA 4
    3. The Role of the FCC and Cell Towers................................. 18
    4. Environmental Consulting Industry and NEPA 25
    5. NEPA Used Excessively in Low-Risk Projects 34
    6. The Hidden Costs to Consumers.. 75
    7. Industry Perspectives 84
    8. A Comparison with Similar Infrastructure 90
    9. Reforming NEPA for the Modern Era................................. 96
    10. Balancing Environmental Protection and Needs 105
    11. Overreach and the Consumer Burden............................... 111
    12. Rethinking NEPA for Telecommunications..................... 124
    13. Restoring NEPA's Purpose.. 132
Bibliography.. 139
Appendices... 142
    Glossary of Terms 143
    Case Studies 144
    Proposed Policy Reforms 145
    Data and Statistics.. 146
About the Author 147

The NEPA Dilemma

Acknowledgements

I am deeply thankful to my wife, whose steadfast support, wisdom, and love have been the bedrock of my life for over 34 extraordinary years. Your encouragement and unwavering belief in me have been my guiding light throughout this process and every step of our shared journey.

To our two children, thank you for filling our lives with immense joy and for teaching us the profound lessons of parenting. Watching you grow has been one of life's greatest privileges, and your achievements continue to inspire and bring us pride.

To our extended family, your support and encouragement have been invaluable throughout this journey. Thank you for standing by us and contributing to the foundation upon which this work was built.

I also extend my heartfelt gratitude to my friends and colleagues. Your insights, shared experiences, and constructive feedback have added depth and dimension to this book. The opportunities to learn from and share in your lives have been incredibly enriching and have greatly influenced the perspectives within these pages.

Lastly, I want to recognize the many individuals I have observed in everyday settings—whether in local malls, theaters, or across the globe—whose interactions and behaviors have served as powerful sources of inspiration. These moments of observation have brought authenticity and real-world relevance to this work.

To all who have been part of this journey, whether directly or indirectly, thank you. Your presence and influence have made this endeavor possible, and I am forever grateful.

The NEPA Dilemma

Douglas B Sims, PhD

Forward

In today's hyper-connected world, telecommunications are as essential as utilities like water and electricity. Yet, while technological advancements have redefined how we work, learn, and communicate, the policies regulating these innovations remain stuck in the past. One glaring example is the misapplication of the National Environmental Policy Act (NEPA)—a 50-year-old law intended to safeguard the environment from large-scale federal projects—now being wielded inappropriately to regulate modern telecommunications technologies. This outdated approach is not only inefficient but also exploitative, as it unjustly inflates consumer expenses each month.

NEPA, enacted in 1970, was designed to assess the environmental impact of significant federal projects like highways, dams, and power plants. However, its use for low-impact telecommunications projects, such as small cell antennas or 5G infrastructure, is a clear mismatch for the needs of today. Consulting firms and industry stakeholders have exploited the law's vague language, stretching its scope to require exhaustive environmental reviews for projects that pose minimal, if any, environmental risk. These unnecessary reviews generate significant profits for a select few while creating delays and inflating costs, with the burden ultimately falling on consumers in the form of higher phone bills.

This misapplication of NEPA stifles progress in deploying critical technologies, forcing American families and businesses to bear the financial burden of inefficiencies they cannot control. Every month, higher connectivity costs are driven not by technological limitations but by a bloated regulatory system designed to protect profits for consulting firms and vested interests. The irony is glaring: a law meant to safeguard the environment is being used as a tool for profiteering while delaying infrastructure improvements that could benefit society at large.

Regulating modern technologies with a policy designed for the 20th century is not just outdated—it's irresponsible. It wastes resources, slows progress, and makes life more expensive for millions of Americans. Regulatory agencies have a duty to adapt to the times, balancing environmental stewardship with the practical realities of modern technology. Continuing to enforce these archaic policies in a fast-evolving digital era is not only inefficient but profoundly unfair to the consumers who bear the cost of this misalignment.

This book calls for a reevaluation of how NEPA is applied to telecommunications. It's not about weakening environmental protections; it's about recalibrating them for today's needs. Policymakers must act to streamline regulations, eliminate redundancies, and ensure that progress comes without unnecessary costs to consumers. By modernizing our regulatory approach, we can achieve a system that supports innovation, reduces costs, and preserves environmental integrity—a win for everyone except those profiting from inefficiency.

Douglas B Sims, PhD

Chapter 1

The NEPA Dilemma in Telecommunications

The National Environmental Policy Act (NEPA), signed into law in 1970, is widely regarded as a cornerstone of environmental protection in the United States. Its primary objective is to ensure that federal agencies carefully consider the environmental consequences of their actions before proceeding with major projects. Central to NEPA are the Environmental Assessments (EAs) and Environmental Impact Statements (EISs), which require federal agencies to assess and disclose the potential impacts of projects ranging from highway construction to energy development (Council on Environmental Quality [CEQ], 2020). NEPA has undeniably played a vital role in preserving the nation's natural resources and ensuring accountability in federally funded projects. However, its application to the rapidly evolving telecommunications industry has exposed significant gaps in its framework, leading to unintended consequences that directly impact consumers.

The telecommunications industry is critical to modern life, powering economic growth, innovation, and connectivity. As society becomes increasingly reliant on mobile technology and high-speed data, the demand for an expansive network of cell towers continues to grow.

These towers are essential for providing reliable service, supporting advancements like 5G networks, and connecting underserved communities in rural areas. To function, however, cell towers must secure licenses from the Federal Communications Commission (FCC). This federal involvement subjects them to NEPA's regulatory requirements, even for projects that have minimal or negligible environmental impact (Federal Communications Commission [FCC], 2022).

While NEPA remains a cornerstone of environmental regulation, its implementation has not evolved to reflect the changing technological and industrial landscape. Initially designed to oversee large-scale projects with significant environmental risks, NEPA now casts a wide net that includes low-impact infrastructure like cell towers. For example, installing a cell tower in an already urbanized area often triggers the same review process as constructing a highway or a dam. This lack of distinction creates inefficiencies and allows for the misapplication of NEPA to projects that do not meaningfully threaten ecosystems or biodiversity.

This misalignment has been exploited by the environmental consulting industry, which leverages NEPA's broad requirements to generate excessive and unnecessary work. By emphasizing lengthy reviews, redundant studies, and bureaucratic processes, consulting firms have created a lucrative business model that benefits them at the expense of project developers and, ultimately, consumers. This dynamic inflates project costs, delays implementation, and drives up phone bills, as telecommunications companies pass compliance costs directly to their customers. As a result, NEPA's outdated framework has inadvertently created an economic burden that stifles the deployment of critical infrastructure and hinders technological progress (Smith, 2019).

Moreover, the outdated scope of NEPA provides a mechanism for stakeholders to impose costly and often unreasonable demands. For instance, Native American tribal groups, while holding an essential role in preserving cultural and historical resources, have increasingly charged exorbitant fees for providing clearance letters for cell tower projects, even in areas where there is no legitimate cultural impact. These inflated fees not only delay projects but also contribute to the rising costs borne by the consumer. This misapplication of NEPA, coupled with the ability of various groups to extract significant payments, further illustrates the inefficiency of the current regulatory framework.

The impact on consumers is significant and often overlooked. The delays and inflated compliance costs resulting from NEPA overreach directly translate into higher phone bills, leaving consumers to pay for inefficiencies in a system designed to protect the environment. For instance, a cell tower project in a suburban area might face months of delays and tens of thousands of dollars in compliance costs, despite posing no discernible environmental threat. These costs, when multiplied across hundreds or thousands of projects nationwide, create a cumulative financial burden that disproportionately affects consumers without delivering meaningful environmental benefits.

NEPA is, without question, a foundational piece of legislation that has advanced environmental protection for decades. However, the rapidly changing landscape of technology and infrastructure demands a reassessment of how it is applied. NEPA was never intended to obstruct essential infrastructure projects like cell towers or to serve as a tool for extracting unwarranted fees. Yet, its current implementation allows these practices to persist, undermining the intent of the law and creating an economic drain on consumers.

This book explores how NEPA's misapplication in the telecommunications industry has led to inefficiencies, inflated costs, and delays that hurt consumers and hinder technological advancement. It examines the broader implications of these practices and argues for reforms that strike a balance between protecting the environment and enabling critical infrastructure development. By adapting NEPA to reflect modern realities, policymakers can reduce unnecessary costs, prevent abuses of the system, and ensure that the law continues to serve its original purpose: safeguarding the environment without stifling progress.

Chapter 2

The Origins and Goals of NEPA

The National Environmental Policy Act (NEPA) of 1969 represents a pivotal moment in the history of environmental legislation, marking the United States' first comprehensive framework for integrating environmental concerns into federal government decisions. NEPA was born from the environmental movement of the late 1960s, a time when the nation grappled with the visible and escalating consequences of rapid industrialization, urban sprawl, and pollution. Environmental disasters, such as the 1969 Santa Barbara oil spill and the burning of the Cuyahoga River in Ohio, starkly highlighted the inadequacies of existing policies to safeguard natural resources and public health. This period of heightened awareness also saw the publication of influential works, like Rachel Carson's *Silent Spring* (1962), which underscored the far-reaching impacts of human activity on ecosystems and biodiversity, galvanizing public demand for environmental reform.

Recognizing the urgent need for a systematic approach to environmental protection, NEPA was drafted to serve as a cornerstone for federal environmental policy. Signed into law by President Richard Nixon on

January 1, 1970, it symbolized bipartisan acknowledgment of the importance of safeguarding the environment for future generations. NEPA's preamble eloquently declared the federal government's commitment "to create and maintain conditions under which man and nature can exist in productive harmony," emphasizing sustainability as a guiding principle for federal actions (Council on Environmental Quality [CEQ], 2020).

At its core, NEPA established a procedural framework requiring federal agencies to assess the environmental impacts of their actions before proceeding. This framework aimed to ensure that government decisions—whether related to infrastructure development, resource extraction, or land use planning—accounted for potential ecological and social consequences. NEPA introduced the Environmental Assessment (EA) and Environmental Impact Statement (EIS) processes, creating a mechanism for identifying environmental risks, exploring alternative approaches, and engaging the public in federal decision-making.

NEPA's passage reflected a broader cultural shift toward environmental stewardship, coinciding with the establishment of the Environmental Protection Agency (EPA) in December 1970 and the celebration of the first Earth Day earlier that year. Together, these milestones signaled a new era in which environmental considerations became integral to public policy. NEPA not only addressed the immediate challenges of unchecked industrial development but also laid the groundwork for future environmental legislation, influencing global approaches to sustainable development and conservation.

Protecting Environmental Quality

One of NEPA's primary goals is to safeguard and improve environmental quality through a structured and transparent decision-making process. Central to this mission is the requirement for federal agencies to evaluate the potential environmental consequences of their actions before they proceed. NEPA mandates that federal projects—ranging from infrastructure development, such as highways and dams, to land use changes and resource extraction activities—be carefully analyzed to identify, mitigate, and, when possible, prevent harm to the environment. This evaluation ensures that environmental considerations are embedded within the planning and execution phases of federally funded or permitted projects, rather than being treated as an afterthought.

The process outlined by NEPA compels agencies to adopt a comprehensive, holistic perspective, examining not only the immediate, localized impacts of a proposed project but also its broader, long-term effects on ecosystems, biodiversity, and public health. For example, when considering a highway construction project, agencies must assess not just the physical alteration of landscapes but also the potential for habitat fragmentation, air and water pollution, and disruptions to wildlife migration patterns. Similarly, resource extraction activities, such as mining or drilling, are scrutinized for their potential to degrade soil quality, contaminate water sources, and contribute to greenhouse gas emissions. This expansive view ensures that decisions are informed by a thorough understanding of how a project might affect interconnected ecological systems over time (U.S. Environmental Protection Agency [EPA], 2023).

NEPA's emphasis on a detailed environmental review reflects a recognition of the complexity and interdependence of natural systems. By requiring agencies to anticipate and address potential harms proactively, NEPA fosters a precautionary approach to federal projects, reducing the likelihood of unintended environmental consequences. Moreover, the law encourages the consideration of alternative approaches that could achieve project goals while imposing less harm. For instance, in a scenario involving the construction of an energy facility, agencies might explore renewable energy options or alternative siting locations to minimize environmental disruption.

Beyond environmental protection, NEPA also underscores the importance of public involvement in federal decision-making. By integrating public comments and expert feedback into the environmental review process, NEPA ensures that diverse perspectives inform project planning. This inclusive approach not only enhances the quality of decision-making but also fosters greater accountability and trust between federal agencies and the communities they serve. In this way, NEPA promotes the dual objectives of environmental stewardship and democratic governance, ensuring that federal actions align with the long-term well-being of both natural ecosystems and human populations.

Ensuring Public Input in Government Projects

NEPA places a strong emphasis on transparency and public involvement, making these principles integral to federal decision-making processes. The law grants citizens the opportunity to review and comment on proposed federal projects through mechanisms such as public hearings, comment periods, and access to environmental documents. These provisions not only enhance the openness of federal planning but also ensure that communities have a voice in shaping projects that may affect their local environments, health, and quality of life. By allowing individuals, organizations, and stakeholders to participate in discussions, NEPA acknowledges the vital role that diverse perspectives play in identifying potential environmental risks and proposing alternative solutions (Council on Environmental Quality [CEQ], 2020).

Public participation under NEPA serves as more than a procedural formality—it reinforces the democratic principle that government actions should align with the needs and concerns of the people they serve. For instance, during the review of a major infrastructure project, community members might highlight localized environmental or cultural considerations that federal agencies might otherwise overlook. These insights can lead to better-informed decisions that reduce harm to vulnerable ecosystems or address potential social inequities. Public input often leads to tangible changes, such as modifications in project design, implementation of additional mitigation measures, or even the selection of entirely different project alternatives.

Moreover, NEPA's emphasis on public involvement ensures accountability and transparency in federal projects. Agencies are required to disclose potential environmental impacts openly and address public concerns within their Environmental Assessments (EAs) or

Environmental Impact Statements (EISs). This fosters trust between federal agencies and the communities they serve, as citizens can see how their input has been considered and incorporated into final decisions. For example, public feedback may uncover environmental justice issues, such as disproportionate impacts on marginalized communities, prompting agencies to revise project plans to address these concerns.

By integrating public engagement into environmental decision-making, NEPA empowers communities to take an active role in shaping the policies and projects that affect their lives. This participatory approach not only enhances the quality and sustainability of federal actions but also strengthens the relationship between government and its citizens, ensuring that environmental governance remains both inclusive and responsive. Through these provisions, NEPA underscores the idea that environmental protection is a shared responsibility, achieved through collaboration between federal agencies and the public they are obligated to serve.

Overview of the Environmental Assessment (EA) and Environmental Impact Statement (EIS) Processes

NEPA's core mechanisms for evaluating environmental impacts are the Environmental Assessment (EA) and the Environmental Impact Statement (EIS), which serve as critical tools for ensuring that federal actions are environmentally responsible. These documents provide a structured, analytical framework to assess the potential effects of proposed projects, offering decision-makers and the public valuable insights into the environmental consequences of federal actions.

The Environmental Assessment (EA) is a concise document used to determine whether a proposed federal action will have significant environmental effects. If the findings indicate no significant impact, a Finding of No Significant Impact (FONSI) is issued, allowing the project to proceed without the need for further analysis. The EA process ensures that smaller-scale projects, which may not necessitate a comprehensive review, are still subject to scrutiny. This step prevents unnecessary delays while maintaining environmental oversight, ensuring that even minor federal actions consider potential environmental impacts (U.S. Environmental Protection Agency [EPA], 2023).

For projects with the potential for significant environmental effects, NEPA requires an Environmental Impact Statement (EIS)—a more detailed and rigorous analysis. The EIS evaluates the scope of a project's environmental impacts, including effects on ecosystems, air and water quality, cultural resources, and public health. It also examines alternatives to the proposed action, offering less harmful or more sustainable options when possible. This in-depth analysis provides a comprehensive overview, helping agencies weigh the environmental trade-offs of their decisions. Additionally, the EIS process requires extensive public input, ensuring that community concerns are incorporated into the final recommendations (Council on Environmental Quality [CEQ], 2020).

Both the EA and EIS processes reinforce NEPA's overarching goals of transparency and accountability. By systematically analyzing potential environmental consequences, these mechanisms ensure that federal agencies make informed decisions that prioritize environmental stewardship. Furthermore, by involving public participation at multiple stages, these evaluations uphold NEPA's commitment to democratic decision-making, ensuring that community voices are heard and addressed.

Together, the EA and EIS serve as the analytical backbone of NEPA's regulatory framework. They not only provide a pathway for minimizing harm to the environment but also foster collaboration among federal agencies, stakeholders, and the public. These mechanisms exemplify NEPA's role as a cornerstone of environmental governance, balancing developmental needs with the imperative to protect the natural world for future generations.

Environmental Assessment (EA)

An Environmental Assessment (EA) serves as a preliminary document within the NEPA framework, designed to evaluate whether a proposed federal action necessitates a more comprehensive Environmental

Impact Statement (EIS). Its purpose is to provide an initial analysis of potential environmental effects, explore alternatives to the proposed action, and identify mitigation measures to address adverse impacts. The EA allows agencies to assess the environmental implications of a project without committing to the time and resource-intensive EIS process unless significant impacts are anticipated.

If the EA determines that a proposed action will not result in significant environmental effects, the agency issues a Finding of No Significant Impact (FONSI). This determination permits the project to move forward without further NEPA review, effectively expediting the process for projects that are unlikely to cause substantial harm to the environment. By enabling agencies to avoid the EIS process for low-impact projects, the EA is designed to streamline NEPA compliance, ensuring that resources are allocated efficiently while still adhering to the act's environmental protection goals (U.S. Environmental Protection Agency [EPA], 2023).

In addition to assessing environmental impacts, the EA also provides an opportunity for public involvement, reinforcing NEPA's commitment to transparency. Agencies may solicit feedback during the EA process, allowing stakeholders and community members to contribute insights or raise concerns about the project. This ensures that even preliminary reviews consider diverse perspectives and account for localized environmental and social considerations.

The EA process embodies NEPA's intent to balance environmental oversight with procedural efficiency. By offering a mechanism to evaluate federal actions that may have minor environmental effects, the EA allows agencies to address potential issues early, mitigate risks, and proceed without unnecessary delay or expense. This step is critical in managing the growing number of federal projects while upholding NEPA's mission to protect environmental quality.

Environmental Impact Statement (EIS)

For projects anticipated to have significant environmental impacts, NEPA mandates the preparation of a comprehensive Environmental

Impact Statement (EIS). This document represents the most detailed level of analysis under NEPA, thoroughly examining the environmental, social, and economic implications of a proposed federal action. The EIS process is designed to ensure that decision-makers fully understand the consequences of their actions and explore viable alternatives to minimize harm.

The EIS evaluates multiple aspects of the proposed project, including its potential impacts on ecosystems, wildlife, water and air quality, cultural resources, and human communities. A critical component of the EIS is the analysis of alternatives to the proposed action, which includes a "no action" alternative that assesses the potential outcomes if the project is not implemented. This comparison ensures that the chosen course of action reflects a balance between development goals and environmental preservation. The document also identifies and proposes measures to mitigate adverse effects, addressing ways to reduce, offset, or prevent environmental damage associated with the project.

The EIS process is characterized by its rigor and inclusivity. It typically begins with scoping meetings, where agencies solicit input from stakeholders to identify the key issues and concerns to be addressed in the analysis. Following the drafting of the EIS, the document undergoes a public review period, during which individuals, communities, organizations, and other stakeholders can provide comments. These comments are considered in the development of the final EIS, ensuring that diverse perspectives are incorporated into the decision-making process.

Once the EIS is finalized, it forms the basis for the agency's Record of Decision (ROD). The ROD outlines whether the project will proceed, the chosen alternative, and any mitigation measures to be implemented. This document signifies the culmination of the EIS process, providing a transparent and reasoned explanation for the agency's decision (Council on Environmental Quality [CEQ], 2020).

The EIS process embodies NEPA's dual goals of environmental protection and public engagement. While it is often resource-intensive and time-consuming, the comprehensive analysis and stakeholder involvement it entails are integral to ensuring that federal projects align with environmental and societal priorities. By requiring an EIS for projects with significant impacts, NEPA enforces accountability and thoroughness in federal decision-making, ensuring that the long-term consequences of development are carefully considered.

How did NEPA become a significant part of Cell Tower Constructions

The integration of the National Environmental Policy Act (NEPA) into cell tower construction arose from the evolving landscape of telecommunications infrastructure and its regulatory oversight. Originally designed to evaluate the environmental impact of large-scale federal projects, NEPA's application to cell towers is rooted in their reliance on Federal Communications Commission (FCC) licensing. The FCC's involvement in granting licenses for the use of radio frequencies brought cell towers under NEPA's jurisdiction, as any federal action, including licensing, triggers NEPA compliance.

Federal Oversight Meets Rapid Telecommunications Growth

The rapid expansion of cellular networks in the 1990s and 2000s brought the construction of cell towers to the forefront of infrastructure development. As cellular service providers sought to establish nationwide coverage, the demand for towers surged. Cell towers, often located in diverse environments—ranging from urban rooftops to rural landscapes—raised concerns about their potential environmental and cultural impacts. These included effects on bird populations, aesthetics, and proximity to historic sites or culturally significant areas.

To address these concerns, the FCC adopted regulations requiring cell tower developers to comply with NEPA. This included conducting Environmental Assessments (EAs) or Environmental Impact Statements (EISs) if significant environmental impacts were anticipated. Additionally, the National Historic Preservation Act (NHPA) and consultation with Native American tribes became integral to the process, further expanding the scope of environmental review for cell towers.

Application to Small-Scale Projects

Although NEPA was originally intended for projects with substantial environmental impacts, its application to cell towers—often minor structures compared to large federal projects—has raised questions about proportionality. Unlike dams, highways, or industrial developments, cell towers are more comparable to streetlights, power poles, or athletic field lights, which generally do not require extensive environmental review. However, because of their association with federally regulated radio frequencies, NEPA compliance became a requirement, even for towers in low-impact areas such as urban settings or on pre-developed land.

The Role of Environmental Consulting

As NEPA became an established part of the cell tower construction process, it created opportunities for environmental consulting firms to carve out a niche. These firms provide services such as impact assessments, historic preservation reviews, and consultations with tribal entities. While such reviews are critical in sensitive environments, the uniform application of NEPA requirements to all cell tower projects—regardless of their actual environmental impact—has been criticized for leading to overreach and unnecessary costs. In some cases, simple tower modifications or upgrades have required lengthy and costly NEPA reviews, slowing down project timelines and inflating expenses.

Consumer Impacts

The costs of compliance with NEPA for cell towers are not absorbed by telecommunications companies alone—they are passed on to consumers through higher service fees. The growing complexity of NEPA reviews, coupled with consultation fees from multiple stakeholders, contributes to delays and financial burdens that ultimately affect consumers. This dynamic has sparked debates over whether NEPA's application to cell towers serves its original purpose or merely fuels an industry of compliance that burdens consumers and slows infrastructure deployment.

Modern Challenges and the Need for Reform

As technology evolves and the demand for 5G networks and small-cell installations grows, the application of NEPA to cell towers has become increasingly contentious. Small cells, essential for 5G, are often installed on existing infrastructure such as utility poles or streetlights, with minimal environmental impact. Yet, they can still trigger NEPA reviews under current regulations, adding time and expense to projects that arguably pose negligible environmental risks.

This misalignment between NEPA's original intent and its application to modern telecommunications projects has prompted calls for regulatory reform. Critics argue that NEPA's broad application to cell towers, particularly small-scale and urban projects, reflects an outdated approach that fails to account for technological advancements and changing infrastructure needs. Proposals for streamlining NEPA compliance, such as categorical exclusions for low-impact projects, aim to strike a balance between environmental stewardship and efficient infrastructure development.

NEPA's role in cell tower construction underscores the complexity of balancing environmental protection with the rapid pace of technological development. While the act's principles remain vital for safeguarding ecosystems and cultural heritage, its application to cell towers has revealed challenges in adapting federal regulations to modern infrastructure needs. Understanding how NEPA became intertwined with cell tower construction is key to addressing these challenges and finding solutions that align with the realities of today's telecommunications landscape.

NEPA's Intended Focus on Federal Projects with Significant Environmental Impacts

NEPA was originally created to address large-scale federal projects with substantial environmental impacts, such as highways, dams, and significant industrial developments. The law aimed to ensure that federal actions were undertaken with a comprehensive understanding of their potential environmental consequences and that alternatives minimizing harm were considered (Smith & Johnson, 2019). This intent was rooted in the broader environmental movement of the 1960s, which sought to balance economic development with environmental stewardship. NEPA provided a framework for federal agencies to evaluate the potential effects of their actions systematically, focusing on projects with clear, significant environmental stakes. Projects with limited or negligible environmental impact were not initially expected to undergo the same rigorous scrutiny.

However, as NEPA's scope expanded over the decades, its application began to include smaller projects, such as cell tower installations and minor infrastructure repairs. This broadening was largely driven by the evolving regulatory landscape and the inclusion of federally linked actions—such as licensing, funding, or permitting—that triggered NEPA compliance. Consequently, projects with minimal environmental risks became subject to extensive reviews, often requiring Environmental Assessments (EAs) or Environmental Impact Statements (EISs). These procedures, originally designed for high-stakes projects, resulted in inefficiencies, delays, and significant financial burdens for smaller-scale initiatives, diverging from NEPA's original purpose.

The application of NEPA to low-impact projects has sparked ongoing debate about its effectiveness and relevance in today's rapidly changing infrastructure and technological context. Critics argue that subjecting small projects to NEPA review creates unnecessary procedural hurdles and costs, diverting attention and resources away from addressing more substantial environmental challenges. For example, requiring full NEPA compliance for minor cell tower modifications in urban areas or already developed settings often results in redundant assessments that yield limited environmental benefits. These critics contend that this expansion undermines NEPA's core mission, shifting its focus from mitigating large-scale environmental risks to enforcing procedural requirements that may add minimal value to environmental protection.

Supporters of NEPA's broad application, however, highlight its foundational role in maintaining accountability and transparency across all federally linked projects, regardless of size. They argue that even seemingly minor projects can have cumulative environmental impacts or raise significant concerns for local communities. NEPA's public engagement requirements provide opportunities for stakeholders to voice concerns and influence decisions, ensuring that federal actions remain accountable to the public (CEQ, 2020). Proponents maintain that this democratic process is essential for upholding environmental protections across all scales of development.

Despite these differing perspectives, NEPA remains a vital tool for safeguarding the environment and fostering public participation in federal decision-making. Its principles of informed decision-making, transparency, and public engagement have shaped national environmental policies and set a global standard for environmental accountability. Yet, the application of NEPA has evolved in ways that were not anticipated at its inception, often encompassing projects with minimal environmental risks. For instance, the use of NEPA reviews in the telecommunications sector, particularly for cell tower construction, has raised questions about whether the law's processes are proportionate to the environmental stakes involved.

As technology advances and infrastructure demands increase, the tension between NEPA's original intent and its modern implementation becomes more pronounced. The rise of small-scale infrastructure, such as 5G networks and microcell installations, underscores the need for a reassessment of NEPA's scope. Policymakers and stakeholders face the challenge of preserving NEPA's effectiveness while reducing inefficiencies that burden low-impact projects. Streamlining NEPA compliance for minor federal actions, perhaps through categorical exclusions or tiered review processes, could help maintain the law's integrity while ensuring it remains relevant in addressing today's environmental and infrastructure challenges.

Understanding NEPA's origins, goals, and processes is crucial for appreciating its transformative impact on federal projects and environmental policy. However, adapting its framework to reflect modern realities is essential for maintaining its balance between protecting the environment and supporting efficient development. A reassessment of NEPA's application to small projects, particularly in sectors like telecommunications, could better align the law with its intended purpose while addressing the inefficiencies that have emerged over time.

Chapter 3

The Role of the FCC and Cell Towers

The Federal Communications Commission (FCC) serves as the cornerstone of telecommunications regulation in the United States, wielding significant authority over the allocation and licensing of radio frequencies. This regulatory oversight ensures the efficient and equitable use of the electromagnetic spectrum, a finite and critical resource for modern communication technologies. However, the FCC's involvement in telecommunications projects triggers compliance with federal laws such as the National Environmental Policy Act (NEPA), often subjecting cell tower installations and modifications to extensive environmental reviews. While the intent is to safeguard environmental and cultural resources, the application of NEPA to low-impact projects, such as small cell installations and colocations, has created regulatory hurdles that slow infrastructure deployment and inflate costs. Understanding the FCC's role in this process is essential to addressing these inefficiencies and streamlining the development of vital telecommunications infrastructure.

The Federal Communications Commission (FCC) and Its Licensing of Radio Frequencies

The Federal Communications Commission (FCC) plays a pivotal role in regulating telecommunications infrastructure, primarily through its authority over licensing radio frequencies. The electromagnetic spectrum—a finite and essential resource—must be carefully managed to ensure that it is allocated and utilized efficiently for various communications purposes, including cellular networks, radio broadcasting, emergency services, and satellite communications (FCC, 2023). The FCC's oversight ensures that competing demands for spectrum access are balanced, enabling seamless connectivity and preventing interference that could disrupt critical communication channels. However, the FCC's licensing process triggers compliance with the National Environmental Policy Act (NEPA), introducing an additional layer of regulatory complexity.

Under NEPA, any federal action—including the FCC's granting of spectrum licenses—requires an evaluation of the potential environmental impacts. This mandate means that projects relying on FCC-approved frequencies, such as the construction of cell towers, must undergo environmental assessments (EA) or more detailed Environmental Impact Statements (EIS) to determine their effect on ecosystems, historic properties, and public health. For instance, a proposed cell tower might be reviewed for its impact on local bird populations, potential interference with historic landmarks, or proximity to sensitive habitats, even when the actual risks are negligible (Council on Environmental Quality [CEQ], 2020). The NEPA review process, though essential for large-scale projects with significant environmental implications, often applies indiscriminately to low-impact telecommunications projects, creating unnecessary delays and expenses.

The intersection of NEPA requirements and FCC licensing illustrates the broader challenge of aligning environmental compliance with technological progress. While the FCC aims to facilitate the rapid deployment of telecommunications infrastructure, particularly for next-generation 5G networks, NEPA reviews can impede these efforts. For example, the Spectrum Frontiers Order, an FCC initiative designed to accelerate 5G rollout, seeks to streamline infrastructure development by reducing regulatory barriers. Yet, the requirement for environmental reviews under NEPA often conflicts with this goal, especially for projects with minimal environmental consequences. This tension reflects the difficulty of balancing the need for robust environmental protections with the demand for modern, high-speed connectivity.

Delays and costs associated with NEPA reviews for telecommunications infrastructure have broader implications for industry stakeholders and consumers. Developers often face prolonged approval processes, which can increase project budgets and timelines. These additional costs are frequently passed down to consumers, contributing to higher costs for telecommunications services. Critics argue that this regulatory overlap represents an inefficiency in the system, as many low-impact projects are subjected to the same rigorous scrutiny intended for large-scale developments with substantial environmental risks.

Moreover, the broad application of NEPA to FCC-licensed projects has sparked debate over whether the regulatory framework has kept pace with advancements in technology and infrastructure deployment. While the FCC is responsible for managing spectrum use, its involvement in environmental compliance highlights a growing need for updated policies that reflect the realities of modern telecommunications. For example, small cell installations, critical for 5G networks, often have a much lower environmental impact than traditional macro towers but are still subjected to NEPA requirements due to FCC licensing. Critics argue that such reviews are disproportionate to the projects' actual risks and call for reforms to streamline processes for low-impact installations while maintaining oversight for projects with significant environmental implications.

In response, there have been calls to modernize NEPA's application to telecommunications infrastructure to reduce redundancies and inefficiencies. Proponents of reform suggest creating exemptions or categorical exclusions for low-impact projects, particularly in urbanized or developed areas where environmental risks are minimal. These adjustments could enable the FCC to focus its regulatory efforts on projects with meaningful environmental considerations while supporting the rapid deployment of critical infrastructure. Such reforms would align with the FCC's broader mission of fostering technological innovation and connectivity while preserving its commitment to environmental stewardship.

The FCC's dual role in spectrum management and environmental compliance underscores the complexities of federal oversight in a rapidly evolving technological landscape. While the agency's regulatory framework is essential for ensuring the equitable and efficient use of the electromagnetic spectrum, its intersection with NEPA highlights the challenges of adapting legacy policies to meet modern needs. As the telecommunications industry continues to expand, particularly with the growing demands of 5G and beyond, balancing environmental protection with infrastructure deployment will remain a critical area of focus.

The Rapid Growth of the Cell Tower Industry to Support 5G and Beyond

The cell tower industry is undergoing unprecedented growth driven by the escalating demand for fast, reliable wireless communication. The shift from 4G LTE to 5G networks has been a significant catalyst, as 5G technology necessitates a much denser infrastructure of cell towers and small cell installations to achieve its promise of faster speeds, lower latency, and expanded connectivity. Unlike previous generations of cellular networks, 5G relies on a web of smaller antennas distributed across urban and suburban landscapes. This growth trajectory is staggering: by 2025, the number of cell towers worldwide is projected to surpass 7 million, fueled by the expansion of 5G infrastructure and its integration with advanced technologies like the Internet of Things (IoT) and autonomous vehicles (GMSA, 2022).

However, the rapid deployment of 5G networks has brought the regulatory landscape, particularly the National Environmental Policy Act (NEPA), into sharp focus. NEPA, originally designed to oversee large-scale federal projects, now applies to projects requiring Federal

Communications Commission (FCC) licensing, including the installation of telecommunications infrastructure. This requirement poses unique challenges for the deployment of 5G infrastructure. While earlier networks relied on a smaller number of large towers to cover expansive areas, 5G requires thousands of smaller, localized installations to ensure consistent coverage and functionality. Despite their minimal environmental footprint, these small cell installations are often subjected to the same rigorous NEPA review processes as large-scale infrastructure projects, leading to inefficiencies and delays.

Critics contend that applying NEPA reviews to small 5G installations creates a disproportionate burden given their negligible environmental impact, especially in urban and suburban areas where existing infrastructure can often be utilized. Many argue that these reviews, which were originally intended for projects with significant potential for environmental harm, slow down the rollout of critical telecommunications infrastructure, increasing costs and impeding technological progress. The procedural requirements of NEPA, including environmental assessments (EA) or more exhaustive environmental impact statements (EIS), can delay projects for months or even years. These delays not only drive up the costs of deployment but also hinder access to the economic and societal benefits of 5G connectivity.

Proponents of NEPA's continued application, however, emphasize the importance of accountability and public input in infrastructure projects. They argue that even small-scale installations can have cumulative effects on sensitive environments, historic sites, or underserved communities. NEPA reviews, they assert, provide a necessary check on projects that could otherwise bypass environmental and community considerations. This perspective highlights the ongoing tension between the need for regulatory oversight and the imperative to accelerate technological innovation.

The current regulatory framework underscores the necessity of modernizing NEPA to accommodate the rapidly evolving telecommunications landscape. Developing streamlined processes for projects with minimal environmental impact, such as categorical exclusions for small cell installations in urban areas, could help balance the goals of environmental protection with the urgent demand for advanced infrastructure. Without such reforms, the bottlenecks created by outdated regulatory processes will continue to impede the efficient deployment of 5G networks, affecting both industry stakeholders and consumers who bear the financial burden of increased costs and delays.

Comparison of Cell Towers with Similar Structures

Cell towers are frequently compared to other familiar structures, such as streetlights, utility poles, and sports field lighting, to underscore the inconsistencies in regulatory oversight and the disproportionate application of the National Environmental Policy Act (NEPA) to telecommunications infrastructure. These parallels reveal a regulatory imbalance, as similarly impactful structures often escape the scrutiny imposed on cell towers.

Streetlights and utility poles are fixtures of urban and rural landscapes, installed routinely without significant environmental oversight. Sports field lighting, comparable in height and visual presence to cell towers, is similarly exempt from the extensive reviews required of telecommunications infrastructure. Despite their analogous nature, cell towers face rigorous NEPA evaluations due to the Federal Communications Commission's (FCC) involvement. Even projects with minimal environmental risks are subjected to this exhaustive process, a requirement not applied to streetlights or utility poles overseen by state or local authorities (FCC, 2023).

The key distinction lies in the federal nexus introduced by FCC licensing, which triggers NEPA compliance for cell towers. By contrast, the installation of streetlights, utility poles, and sports field lighting typically falls under state or local jurisdiction, circumventing federal environmental mandates. This regulatory discrepancy raises questions about whether applying NEPA to cell towers—particularly in urban or already-developed areas with minimal environmental impacts—is justified.

Case studies reveal the practical consequences of this uneven regulatory landscape. NEPA reviews for cell towers often result in redundant assessments, unnecessary delays, and inflated costs. These expenses are frequently passed on to consumers in the form of higher telecommunications bills. For example, the rollout of small cell infrastructure essential for 5G networks has been significantly delayed in several cities due to required environmental reviews. These installations, which have minimal environmental impact and are often attached to existing urban structures, face procedural hurdles that comparable projects—like streetlight upgrades—avoid altogether.

By contrast, streetlight or sports field lighting projects proceed without comparable scrutiny, even when their visual and environmental footprints are on par with those of cell towers. This discrepancy highlights inefficiencies in applying NEPA uniformly across federally linked projects. The case of cell towers illustrates how outdated regulatory frameworks can hinder infrastructure development, imposing unnecessary costs while yielding limited environmental benefits. Revisiting the application of NEPA to ensure it aligns with the evolving landscape of infrastructure projects is essential to promote efficiency, equity, and consumer affordability.

Douglas B Sims, PhD

Chapter 4

Environmental Consulting Industry and NEPA

The expansion of NEPA's scope has catalyzed the growth of an extensive environmental consulting industry, particularly in sectors such as telecommunications infrastructure. Originally intended to address large-scale federal projects with significant environmental impacts, NEPA's application has broadened to encompass a wide range of projects, including those with minimal environmental risks, such as cell tower installations. This shift has created an environment where environmental consulting firms play a central role in ensuring compliance with NEPA's complex regulatory requirements. These firms offer specialized expertise in preparing Environmental Assessments (EAs), Environmental Impact Statements (EISs), and other compliance documents, which are now standard in many infrastructure projects.

The construction and maintenance of cell towers, especially in the era of 5G technology, have become a particularly lucrative niche for these firms. The rapid expansion of 5G networks requires the deployment of thousands of small cell installations and new tower sites to meet growing connectivity demands. Each of these projects involves FCC licensing, which triggers NEPA compliance requirements. Consequently, consulting firms are frequently contracted to assess the potential environmental, cultural, and historical impacts of these projects, even in cases where the risk of significant harm is negligible. This proliferation of reviews has turned cell tower projects into a reliable revenue stream for the environmental consulting industry.

What makes this industry particularly noteworthy is how NEPA reviews for cell towers have become standardized and routine, regardless of the project's environmental complexity or impact. Many reviews follow formulaic processes, using templates and checklists that prioritize procedural compliance over meaningful environmental assessment. This standardization streamlines the workload for consulting firms but often results in redundant or unnecessary reviews, particularly for projects in urbanized or low-risk areas. For example, small cell installations on pre-existing utility poles or minor upgrades to existing towers frequently undergo the same level of scrutiny as entirely new developments in environmentally sensitive areas.

The chapter delves into specific examples that illustrate how NEPA reviews have become excessive and unnecessary in certain contexts. Case studies highlight instances where environmental reviews delayed low-impact projects, such as cell towers installed in agricultural fields or urban locations with minimal ecological sensitivity. These reviews, while technically compliant with NEPA, provide little environmental value but significantly increase project timelines and costs. This inefficiency not only slows the deployment of critical telecommunications infrastructure but also burdens the industry with unnecessary expenses.

Finally, the chapter examines the financial and temporal burdens these processes impose on the telecommunications sector and consumers. The costs associated with NEPA reviews are often substantial, with consulting fees, administrative expenses, and project delays driving up overall expenditures. These costs are frequently passed on to consumers in the form of higher phone bills and service fees. Moreover, the time delays associated with compliance reviews hinder the timely deployment of advanced technologies like 5G, which rely on dense networks of infrastructure to deliver faster speeds and better connectivity. The chapter argues that reforming NEPA's application to focus on high-risk projects could alleviate these burdens, ensuring that environmental reviews provide meaningful protection without stifling technological progress.

The Emergence of Environmental Consulting Firms Specializing in NEPA Compliance for Cell Towers

Environmental consulting firms began proliferating in the 1970s, shortly after the enactment of the National Environmental Policy Act (NEPA). Initially, these firms concentrated on large-scale federal projects that fell squarely within NEPA's intended scope, such as highway construction, dam building, and energy developments. These projects, due to their significant environmental impact, required thorough Environmental Assessments (EAs) or Environmental Impact Statements (EISs) to ensure compliance with NEPA's goals of minimizing harm to natural ecosystems and communities. However, as NEPA's application expanded over the decades to include smaller projects, such as cell tower construction, the consulting industry adapted and flourished to meet the rising demand for compliance services. This evolution reflected a shift in NEPA's practical implementation, where even low-risk projects became subject to its requirements. Telecommunications projects, in particular, emerged as a lucrative niche because they involve Federal Communications Commission (FCC) licensing, which automatically triggers NEPA reviews (FCC, 2023).

The role of environmental consulting firms in telecommunications projects is to navigate the regulatory complexities of NEPA. This often involves conducting EAs to determine whether a more detailed EIS is necessary. For cell tower projects, consultants evaluate a range of potential environmental and community impacts, including effects on local ecosystems, wildlife habitats, cultural and historic preservation, and even public health concerns related to radiofrequency emissions. While these assessments are vital for projects located in sensitive or undeveloped areas, the process has increasingly been applied to low-risk installations, such as upgrades to existing towers or small cell sites mounted on pre-existing infrastructure like utility poles.

The consulting industry has effectively commoditized the NEPA review process, transforming what was initially intended as a thorough and meaningful evaluation into a standardized bureaucratic exercise. This commoditization often prioritizes procedural compliance over substantive environmental protection. For example, consulting firms frequently use template-driven assessments and checklists that focus on meeting regulatory requirements rather than conducting in-depth analyses of potential environmental consequences. This shift has allowed firms to scale their operations and handle a high volume of low-risk projects, turning NEPA compliance into a routine but profitable venture (Smith & Johnson, 2020).

The commoditization of NEPA reviews has had far-reaching implications. On one hand, it has fueled the growth of the environmental consulting industry, creating thousands of jobs and fostering expertise in environmental compliance. On the other hand, it has diluted the original intent of NEPA by diverting attention and resources from genuinely high-impact projects to minor ones. For instance, a minor modification to an existing cell tower or the installation of a small cell antenna may undergo the same level of scrutiny as a new tower built in an ecologically sensitive area. This approach inflates the costs and timelines of even the simplest projects, creating inefficiencies that ripple through the telecommunications sector and ultimately affect consumers in the form of higher service costs.

Furthermore, as consulting firms became integral to NEPA compliance, their influence grew within the regulatory ecosystem. Their expertise and established relationships with federal agencies often make them indispensable to project developers. This dependency, while practical, raises concerns about the impartiality of the process, as firms benefit financially from perpetuating the need for exhaustive reviews, even when the environmental risks are minimal. The growing reliance on these firms highlights the tension between NEPA's noble environmental goals and its bureaucratic realities, emphasizing the need for a recalibration of its scope to align with modern project demands and environmental priorities.

In summary, the proliferation of environmental consulting firms was a direct response to NEPA's expansive reach. While these firms play a crucial role in facilitating compliance, their commoditization of the process for low-impact projects such as cell towers raises questions about the efficiency and fairness of NEPA's current application. As the demand for telecommunications infrastructure continues to grow, particularly with the rollout of 5G networks, the balance between meaningful environmental protection and regulatory efficiency becomes more critical than ever.

How NEPA Reviews for Cell Towers Have Become a Routine Process

The application of NEPA to telecommunications infrastructure has become increasingly standardized, applying the same rigorous processes to cell towers regardless of their environmental impact. This approach means that small-scale projects, such as the installation of small cell antennas on existing utility poles or minor upgrades to previously approved towers, are often subjected to the same NEPA procedures as new, large-scale towers constructed in environmentally sensitive areas. These reviews typically involve conducting environmental site assessments, consulting with stakeholders, and preparing detailed documentation, even when the potential for significant environmental impact is negligible (EPA, 2023). As a result, projects with minimal environmental risks often face disproportionate regulatory scrutiny, leading to inefficiencies and inflated costs.

This standardized approach largely stems from a culture of regulatory caution. Federal agencies and developers often prioritize comprehensive compliance to avoid the potential for litigation or project delays. NEPA's citizen suit provision allows the public to challenge projects on environmental grounds, incentivizing agencies and developers to adopt a "better safe than sorry" mentality. This precautionary approach ensures that all procedural boxes are checked, even for projects that pose little or no environmental risk. While this thoroughness reduces legal vulnerabilities, it also imposes significant administrative burdens on projects that may not warrant such detailed review.

Environmental consulting firms have played a central role in facilitating this standardization. To meet the demand for NEPA compliance, many firms have developed streamlined processes to handle routine reviews efficiently. These processes often rely on templates, checklists, and standardized methodologies designed to meet regulatory requirements with minimal customization. While these tools expedite the compliance process, they frequently provide little substantive analysis of actual environmental risks. For example, a minor modification to an existing cell tower, such as upgrading its radio equipment, might trigger a NEPA review that evaluates wildlife impacts, cultural resource concerns, and public health considerations—issues unlikely to arise in such a low-impact scenario (CEQ, 2020). This approach reduces the depth and utility of NEPA reviews, transforming them into procedural formalities rather than meaningful evaluations of environmental consequences.

The reliance on standardized reviews also perpetuates a cycle of redundant assessments. In many cases, telecommunications projects involve infrastructure that has already undergone previous NEPA reviews. For instance, adding small cell equipment to a utility pole in a densely populated urban area, where environmental risks are minimal and well understood, may still require a full review process. These repeat assessments often provide little new information but still require time, effort, and financial resources. The result is a system where NEPA compliance becomes a box-checking exercise rather than a strategic tool for environmental protection.

This routine application of NEPA has significant financial and temporal consequences for the telecommunications industry. The costs associated with redundant reviews are ultimately passed on to consumers, contributing to higher phone bills and slower deployment of critical infrastructure, such as 5G networks. Moreover, the time required to complete these reviews delays project timelines, hindering the industry's ability to meet growing demand for faster and more reliable wireless connectivity. These inefficiencies highlight the need for a more nuanced application of NEPA that distinguishes between high-impact projects requiring rigorous review and low-impact projects that could benefit from streamlined procedures.

While the intent of NEPA is to safeguard environmental quality and ensure public involvement, its current implementation for telecommunications infrastructure raises questions about proportionality and effectiveness. By treating all projects with the same level of scrutiny, regardless of their actual risks, the NEPA process risks undermining its credibility and burdening industries with unnecessary costs. A reevaluation of how NEPA is applied to low-impact telecommunications projects could help restore its original purpose—protecting the environment from significant harm—while allowing for more efficient infrastructure development.

Case Studies of Unnecessary Environmental Reviews for Low-Impact Sites

Several case studies underscore the inefficiencies and misapplications of NEPA in the telecommunications sector, particularly for low-impact projects. In one example, the installation of small cell antennas on pre-existing utility poles in a densely populated urban area faced significant delays due to the NEPA review process. Despite the urban setting and the use of existing infrastructure, the project was delayed for over six months as the required Environmental Assessment (EA) and public comment periods were completed. The assessment ultimately concluded there were no significant environmental impacts. However, the mandatory compliance process led to increased costs and project delays, providing little environmental value in return (FCC, 2023).

Another case involved the proposed construction of a cell tower in a rural agricultural field that had already been cleared of vegetation. Although the site presented no environmental risks, a full EA was required, adding several months to the project timeline and substantially increasing costs. The review included analyses of potential impacts on wildlife, cultural resources, and public health—concerns that were irrelevant given the context of the cleared, low-risk site. The additional time and expense provided no measurable benefit to environmental protection but significantly burdened the project's timeline and budget.

Similarly, routine upgrades to existing towers often face unnecessary NEPA reviews. For example, adding new antennas or replacing equipment on a previously approved tower with no changes to the site's footprint or environmental context still requires a full review. These redundant assessments often revisit issues already addressed in earlier evaluations, further inflating costs and delaying project completion (Smith & Johnson, 2020). In many cases, the outcomes of these reviews merely confirm the absence of significant environmental impact, raising questions about the efficiency and necessity of applying NEPA so broadly.

These examples illustrate how NEPA, originally intended to safeguard the environment from substantial harm, is often misapplied to low-impact telecommunications projects. The delays and costs associated with redundant reviews not only hinder infrastructure development but also highlight the need for a more streamlined approach to NEPA compliance for projects with minimal environmental risks. A reevaluation of these processes could reduce inefficiencies while maintaining NEPA's core principles of environmental protection and public involvement.

Analysis of the Financial and Time Costs Associated with These Reviews

The financial burden imposed by NEPA reviews is considerable, particularly in the telecommunications sector. Environmental consulting firms often charge significant fees for preparing Environmental Assessments (EAs) and Environmental Impact Statements (EISs), even for projects that pose minimal environmental risks. These costs, which can reach tens of thousands of dollars for a single review, become a direct expense for telecommunications companies. Ultimately, these expenses are transferred to consumers through higher service fees and increased phone bills, creating an economic ripple effect that disproportionately impacts the public (CEQ, 2020).

The delays associated with NEPA compliance further exacerbate these financial burdens. Completing NEPA reviews often requires months of procedural steps, including environmental site evaluations, stakeholder consultations, and public comment periods. These time-consuming processes slow the deployment of critical telecommunications infrastructure, such as 5G networks. Unlike earlier networks, 5G technology relies on dense, small-cell installations that must be widely distributed for optimal coverage. The requirement for NEPA reviews, even for minor installations, creates significant barriers to infrastructure expansion, delaying access to improved connectivity and technological advancements.

Critics of the current application of NEPA argue that the costs and delays are disproportionate to the environmental risks posed by most telecommunications projects. The uniform application of NEPA to all federally linked projects, irrespective of their environmental impact, has diverted the law from its original intent of addressing significant risks. This blanket approach often results in redundant reviews and procedural inefficiencies, channeling resources toward low-impact projects while neglecting high-risk developments that merit closer scrutiny. For example, a minor modification to an existing cell tower can trigger the same level of review as a new installation in an environmentally sensitive area, despite the stark difference in potential impact.

Reforming NEPA to better align with its original purpose could alleviate these challenges. By narrowing its focus to genuinely high-risk projects, NEPA reviews could prioritize meaningful environmental protections while reducing unnecessary financial and temporal burdens. Streamlining the process for low-impact projects would allow agencies and consulting firms to allocate resources more effectively, ensuring that environmental reviews address substantive concerns rather than procedural obligations. Such reforms could also accelerate infrastructure deployment, reduce consumer costs, and reinforce NEPA's role as a tool for safeguarding the environment without stifling technological progress.

Chapter 5

How NEPA Has Been Used Excessively in Low-Risk Projects

The National Environmental Policy Act (NEPA), originally designed to safeguard ecosystems from significant federal projects, has expanded far beyond its initial purpose. In the telecommunications sector, its application to cell tower projects—ranging from small antenna upgrades to new installations in urban areas—has created a maze of regulatory requirements that often deliver little environmental benefit. What was once a tool for evaluating major infrastructure projects like highways and dams has become a burdensome, costly process for minor, low-impact projects. This overreach has allowed environmental consulting firms to capitalize on vague NEPA guidelines, while excessive fees from tribal consultations and redundant reviews inflate project costs. These inefficiencies not only delay the deployment of critical telecommunications infrastructure, such as 5G networks, but also pass unnecessary costs onto consumers, ultimately driving up phone bills. This chapter examines the roots of NEPA overreach in the cell tower industry, exploring how regulatory processes and stakeholder interests contribute to inflated project budgets, delayed timelines, and a telecommunications system hampered by outdated oversight.

Reviews for Towers in Already Urbanized or Low-Impact Areas

NEPA was established to address significant environmental risks posed by large-scale federal projects, such as highways, dams, and industrial developments, where the potential for harm to ecosystems, biodiversity, and public health warranted thorough evaluation. However, the law's application has expanded to encompass low-risk projects, including those in urbanized areas dominated by pre-existing infrastructure. Projects such as small cell installations for 5G networks or minor upgrades to existing towers—actions that typically involve minimal physical alterations—are now often subject to the same NEPA requirements as large-scale developments in pristine or ecologically sensitive areas.

For example, a small antenna upgrade on a utility pole in a densely populated urban area can trigger an Environmental Assessment (EA) or, in some cases, an Environmental Impact Statement (EIS). These evaluations involve extensive documentation, stakeholder consultations, and public review periods, consuming time and financial resources that often outweigh the environmental stakes. Despite the rigorous process, such reviews in urban areas rarely uncover significant risks because the landscapes are already heavily modified, and the incremental impacts of small projects are negligible (Smith & Johnson, 2020). Similarly, cell towers placed on rooftops or within commercial zones must often undergo the same level of scrutiny as projects in undeveloped regions, despite having no direct interaction with natural ecosystems or endangered species.

This blanket application of NEPA creates inefficiencies, delays infrastructure deployment, and inflates project costs. For instance, small cell installations—which are critical for expanding 5G networks—often face delays of several months due to mandatory NEPA reviews, even when they involve minimal environmental impact. These delays not only slow the rollout of advanced telecommunications technologies but also drive up costs for telecommunications companies, which are passed on to consumers through higher service fees. The added financial burden stems from the requirement to hire environmental consultants, complete redundant studies, and navigate regulatory red tape, all for projects that contribute little to environmental degradation.

Critics argue that this overly cautious approach undermines NEPA's original intent, diverting resources from projects with genuine environmental risks to low-impact activities that require only procedural compliance. While ensuring accountability and oversight is important, applying NEPA uniformly to all federally linked projects—regardless of their scale or impact—dilutes the law's effectiveness. Reforming NEPA to focus on high-risk projects, particularly those in ecologically sensitive or undeveloped areas, could streamline compliance, reduce unnecessary costs, and enable more efficient deployment of critical infrastructure without compromising environmental protection.

Redundant Assessments for Modifications to Existing Infrastructure

A recurring issue in the telecommunications sector is the requirement for NEPA reviews on minor modifications to existing infrastructure, such as adding antennas, upgrading equipment, or performing routine maintenance. These modifications generally result in negligible environmental changes, yet they still trigger extensive compliance processes under NEPA. The law mandates Environmental Assessments (EAs) or, in some cases, Environmental Impact Statements (EISs) for federally linked projects, including those regulated by the Federal Communications Commission (FCC). Even when a cell tower has previously undergone environmental reviews, minor updates to its equipment or operations can necessitate an entirely new round of assessments. This redundancy occurs despite the lack of significant changes to the tower's footprint, operational impact, or surrounding environment.

For instance, adding an antenna to a pre-existing tower—essentially swapping out or enhancing current capabilities—can prompt another round of environmental scrutiny, including stakeholder consultations, public comment periods, and the generation of lengthy documentation. This requirement exists even if the original tower's placement, structural impact, and ecological footprint have already been thoroughly evaluated and deemed negligible. The same holds true for replacing outdated equipment with more energy-efficient or technologically advanced alternatives, which could ostensibly reduce environmental impact. Instead of facilitating such upgrades, NEPA's blanket application imposes procedural hurdles that neither enhance environmental protection nor acknowledge the realities of modern telecommunications infrastructure (FCC, 2023).

The burden of these redundant reviews is both financial and operational. Telecommunications providers are required to allocate substantial resources to comply with these requirements, including hiring environmental consultants to perform assessments and complete the associated paperwork. These costs quickly accumulate, often reaching tens of thousands of dollars per project, even for upgrades that are effectively maintenance-level changes. Moreover, the additional layers of compliance result in significant delays, potentially postponing project completion by several months. Such delays are particularly problematic given the accelerated timeline for deploying 5G networks, which require a dense network of small cells and upgraded towers to meet the demand for faster, more reliable wireless connectivity.

The inefficiency of this process extends beyond direct costs and delays, impacting broader telecommunications goals. With consumer demand for advanced network capabilities increasing, the need for a streamlined regulatory process is critical to ensure timely upgrades and expansions. Yet, redundant NEPA reviews slow progress and inflate budgets, hindering efforts to modernize infrastructure and expand network coverage to underserved areas. Critics argue that applying NEPA to minor modifications undermines its original purpose by diverting resources away from meaningful environmental oversight and toward procedural formalities. Reforming NEPA to exempt low-impact modifications or to streamline reviews for pre-assessed sites could significantly reduce these inefficiencies, enabling telecommunications providers to focus on innovation and service delivery while still adhering to environmental safeguards.

How Consulting Firms Exploit Vague NEPA Guidelines to Generate Unnecessary Reviews

The ambiguity in NEPA's guidelines creates significant opportunities for consulting firms to mandate reviews for projects with minimal environmental risks, driving unnecessary costs and delays. The law's lack of clearly defined thresholds for determining what constitutes a "significant impact" enables consulting firms to interpret the requirements broadly. This flexibility often leads to Environmental Assessments (EAs) being recommended for low-stakes projects that pose negligible environmental concerns. For instance, a routine maintenance task such as upgrading an antenna on an existing cell tower might be flagged as requiring a comprehensive cultural, ecological, and historical review, even when the site has already undergone prior assessments and remains unchanged in terms of its footprint or operations (CEQ, 2020).

This overreach allows consulting firms to secure steady revenue streams by framing even minor modifications as needing extensive reviews. These reviews frequently include tasks such as stakeholder consultations, field surveys, and voluminous documentation—processes that are time-consuming and costly but rarely reveal new or meaningful environmental risks. In urban areas, for example, where infrastructure such as utility poles and cell towers is commonplace, firms may still advocate for exhaustive evaluations, despite the low likelihood of significant environmental or cultural impacts.

The practice is not limited to urban projects; even in rural or industrial settings, consulting firms often treat routine activities, such as replacing outdated equipment, as opportunities to justify comprehensive reviews. This approach benefits the firms financially but often does little to further NEPA's original goal of protecting critical ecosystems and addressing genuine environmental threats. Instead, it shifts the focus of NEPA compliance from substantive environmental oversight to procedural bureaucracy, creating inefficiencies that ripple across the telecommunications industry. By exploiting these ambiguities, consulting firms contribute to inflated project costs and delayed infrastructure deployment, ultimately passing these burdens onto consumers in the form of higher service fees (Smith & Johnson, 2020).

The lack of standardized criteria for NEPA compliance also allows consulting firms to duplicate efforts unnecessarily. For example, a project that has already undergone thorough environmental review for its initial construction may be subjected to additional EAs for minor changes, despite no significant alteration to the project's scope or impact. Such practices not only inflate costs but also divert resources away from projects with genuine environmental risks, undermining NEPA's intent to focus on meaningful environmental protection. Addressing this exploitation requires clearer guidelines, more consistent application of thresholds for "significant impact," and oversight to prevent redundant and low-value reviews from becoming a routine aspect of NEPA compliance.

Overlapping Studies and Duplicative Assessments That Add Little Value

Consulting firms frequently engage in overlapping and redundant studies for the same project, often requiring separate reviews under multiple regulatory frameworks that could otherwise be consolidated. These duplicative efforts significantly inflate project budgets and extend timelines, creating inefficiencies that provide little or no additional environmental value. For instance, a telecommunications project might necessitate both environmental and cultural assessments, yet these reviews are often conducted independently rather than through a unified and streamlined process. This fragmentation requires additional fieldwork, documentation, and stakeholder consultations, which not only drive up costs but also delay project completion (Smith & Johnson, 2020).

A prime example of this redundancy can be seen in projects involving the construction or modification of cell towers. In such cases, an environmental assessment might examine impacts on local ecosystems, while a separate cultural review evaluates potential effects on historically significant sites. Despite significant overlap in the data collected and the stakeholders consulted, these processes are often treated as distinct obligations, necessitating separate teams, reports, and regulatory filings. This siloed approach results in higher consulting fees and increased administrative overhead, placing a heavy burden on telecommunications providers.

The inefficiency is further exacerbated by inconsistent coordination between agencies. Federal, state, and local regulations sometimes overlap, yet consulting firms rarely integrate these requirements into a single, cohesive review process. For example, a cell tower project might require compliance with both NEPA and state environmental policies, with consulting firms often opting to conduct separate assessments for each jurisdiction. This practice not only duplicates effort but also increases the likelihood of conflicting recommendations, leading to further delays as discrepancies are resolved.

These inefficiencies have a direct financial impact on the telecommunications sector, with consulting fees for redundant studies often reaching tens of thousands of dollars per project. For smaller installations, such as small cell deployments for 5G networks, these costs can represent a significant portion of the total project budget. The delays associated with redundant reviews also slow the rollout of critical infrastructure, particularly in underserved areas where improved connectivity is most needed. Ultimately, these costs are passed on to consumers in the form of higher service fees, making telecommunications services more expensive for end users.

To address these inefficiencies, experts have called for greater integration of regulatory requirements and streamlined review processes. Consolidating environmental and cultural assessments into a unified approach could significantly reduce costs and timelines while maintaining rigorous oversight. Additionally, increased use of categorical exclusions for low-impact projects could help prevent redundant reviews, allowing resources to be focused on projects with genuinely significant environmental or cultural risks. By addressing the inefficiencies associated with duplicative reviews, stakeholders can ensure that NEPA compliance serves its intended purpose without imposing unnecessary burdens on the telecommunications industry and consumers.

The Role of Native American Tribes in the NEPA Process

Tribal involvement in the NEPA process is rooted in the federal government's trust responsibility to Native American tribes, as well as legal obligations to consult on projects with potential impacts on tribal lands, cultural sites, or sacred areas. This obligation is reinforced by Section 106 of the National Historic Preservation Act (NHPA), which requires federal agencies to consider the effects of their actions on historic properties. When telecommunications projects trigger NEPA compliance, particularly through FCC licensing, these consultations are essential for ensuring that tribal cultural and historical sites are protected. For tribes, this process is a critical mechanism for preserving their heritage, as cell tower installations and other infrastructure projects have the potential to disturb culturally significant landscapes, burial grounds, and sacred sites (FCC, 2023).

The integration of tribal consultation into the NEPA process highlights the intersection of environmental and cultural preservation. Section 106 reviews often run parallel to NEPA reviews, with tribes playing a key role in identifying sites that may be impacted by proposed projects. These consultations ensure that federal actions take into account tribal concerns, fostering a collaborative approach to decision-making. However, the coordination of these efforts can be challenging, particularly when multiple tribes claim cultural ties to the same project area. This dynamic often results in prolonged review periods as telecommunications companies navigate diverse tribal feedback and negotiate resolutions to concerns.

The requirement for tribal consultation, while essential for cultural preservation, has also introduced significant complexities for telecommunications companies. Projects often involve outreach to multiple tribes, each with distinct protocols, timelines, and expectations. This multi-tribe engagement increases administrative burdens and creates opportunities for delays, especially when overlapping regulatory frameworks, such as Section 106 and NEPA, require separate yet related reviews. Furthermore, the lack of a centralized system for managing tribal consultations often leads to inconsistent communication and inefficiencies, adding to the challenges faced by companies attempting to comply with federal regulations.

By ensuring that tribes have a voice in the NEPA process, federal policies uphold the importance of cultural and historical preservation. However, the complexities inherent in tribal consultations, coupled with the evolving demands of infrastructure development, highlight the need for streamlined processes that maintain the integrity of tribal input while addressing inefficiencies in the current system.

Charging Exorbitant Fees for Clearance Letters

One of the most contentious aspects of tribal involvement in the NEPA process is the cost of obtaining clearance letters for telecommunications projects. These letters, which verify that a project does not impact tribal cultural or historical resources, have become a significant financial burden for companies. In many cases, tribes charge thousands of dollars per letter, even for projects with negligible environmental or cultural risks. For instance, a simple antenna replacement on an existing tower may still require clearance letters from multiple tribes, each demanding separate fees, despite the minimal likelihood of cultural impact (Smith & Johnson, 2020).

The issue of exorbitant fees has escalated as the number of telecommunications projects requiring NEPA reviews has grown, particularly with the expansion of 5G infrastructure. Some tribes have established fixed fee structures, while others negotiate costs on a case-by-case basis, leading to wide disparities in charges. In extreme cases, fees for a single clearance letter have exceeded tens of thousands of dollars, creating significant financial obstacles for telecommunications companies. These costs are often passed on to consumers through higher phone bills and service fees, highlighting the broader economic impact of inflated clearance letter charges.

The repetitive nature of some clearance letter requests further compounds the issue. For example, if a telecommunications company seeks to upgrade equipment on an existing tower previously reviewed and cleared, it may still need to obtain new clearance letters from the same tribes. This redundancy adds unnecessary costs and delays, as tribes charge fees for what is essentially a reauthorization of previously approved work. Consulting firms involved in facilitating these processes often exacerbate the problem, charging additional fees for managing negotiations and paperwork, further inflating project budgets.

While the fees associated with tribal clearance letters are intended to support cultural preservation efforts, the lack of standardized pricing and the potential for exploitation have raised concerns among stakeholders. Telecommunications companies argue that these costs are disproportionate to the actual risks posed by most projects, particularly those in urban or already developed areas. As the NEPA process continues to evolve, addressing the issue of exorbitant fees is critical to balancing the need for cultural preservation with the economic realities of modern infrastructure development.

Examination of the Role Native American Groups Play in Providing Cultural and Historical Clearances

Federal law mandates that Native American tribes be consulted on projects involving federal licenses, such as those regulated by the Federal Communications Commission (FCC). This obligation arises from the federal government's trust responsibility and legal requirements under the National Historic Preservation Act (NHPA), specifically Section 106, which ensures that projects consider the effects on properties of historical or cultural significance. In the context of NEPA compliance, this means that telecommunications projects, including the construction and modification of cell towers, must engage with tribes to assess potential impacts on sacred sites, burial grounds, and other culturally important areas (FCC, 2023).

Tribal consultations typically result in clearance letters, which certify that a proposed project will not disturb culturally significant sites. These letters are essential for projects to proceed, as they confirm that the tribal concerns have been addressed and that the integrity of historical and cultural resources is safeguarded. This process is crucial for preserving tribal heritage, particularly as modern infrastructure development increasingly encroaches on lands and sites of historical importance. For many tribes, the clearance process serves as a vital mechanism to ensure their voices are heard and their cultural legacy is protected for future generations.

However, the implementation of this requirement often introduces significant financial and logistical challenges. Telecommunications companies are frequently required to engage with multiple tribes, each with its own protocols, timelines, and fees. Clearance letters, while a necessary part of the consultation process, can carry substantial costs. In some cases, tribes charge thousands of dollars for a single letter, even for projects in urbanized or previously assessed areas where the likelihood of cultural disruption is minimal. These fees are intended to support tribal governments and cultural preservation programs but can create significant financial burdens for companies attempting to deploy critical infrastructure (Smith & Johnson, 2020).

The costs associated with clearance letters are compounded by the growing demand for consultations as telecommunications infrastructure expands, particularly with the rollout of 5G networks. Even projects involving minor modifications, such as adding antennas to existing towers, may require new clearance letters, further driving up expenses. This dynamic has raised concerns about the lack of standardized pricing and the potential for exploitation within the consultation process, as fees vary widely between tribes and often lack transparency.

While the consultation process is a cornerstone of protecting tribal heritage, its financial and procedural impacts underscore the need for a balanced approach. Ensuring that tribes are fairly compensated for their expertise and efforts must be weighed against the broader economic implications of high fees, particularly when these costs are ultimately passed on to consumers in the form of higher phone bills. Addressing these challenges requires a more efficient and equitable framework for tribal consultations that preserves cultural resources while minimizing unnecessary financial burdens on infrastructure development.

Instances Where Clearance Letters Become Costly Obstacles

In some cases, tribes charge thousands of dollars for clearance letters, even for projects located in areas where there is little to no historical or cultural significance. These fees, while initially intended to cover administrative costs and compensate tribes for their time and expertise in reviewing projects, have, in certain instances, escalated disproportionately. For larger projects or those requiring consultations with multiple tribes, the cumulative costs can reach tens of thousands of dollars, creating a substantial financial burden for telecommunications companies.

For example, a relatively straightforward cell tower installation in a rural or urban area may trigger consultations with multiple tribes if the project is within the radius of territories historically associated with them. Each tribe may charge separate fees for conducting reviews and issuing clearance letters, even when the project site has been previously assessed and deemed to hold no cultural or historical value. The fees often include expenses for administrative processes, research, and staff time, but a lack of standardized pricing or oversight can lead to wide variations in cost, with some tribes charging significantly higher amounts than others for similar services (FCC, 2023).

These escalating costs are not limited to new tower installations. Routine upgrades or minor modifications to existing infrastructure—such as adding antennas or replacing outdated equipment—may also necessitate new clearance letters, regardless of whether the modifications alter the project's footprint or impact. This recurring requirement for clearance letters drives up project expenses and delays deployment timelines. For telecommunications companies tasked with deploying critical infrastructure, particularly for 5G networks, these fees represent an increasingly significant challenge.

The financial implications extend beyond telecommunications providers, as these added costs are often passed on to consumers in the form of higher service fees. As the demand for faster and more reliable wireless networks grows, the cumulative effect of excessive clearance fees becomes a barrier not only to efficient infrastructure deployment but also to maintaining affordable connectivity for the public. While it is essential to respect and preserve tribal heritage, the current system's inefficiencies and lack of cost regulation highlight the need for a more equitable and transparent approach to balancing cultural preservation with technological progress (Smith & Johnson, 2020).

Documentation of Fees Reaching Thousands of Dollars Per Project

There are documented cases where multiple tribes charge clearance fees for the same project, particularly when the project spans or is located near areas of overlapping tribal interest or jurisdiction. This scenario is common in telecommunications projects, such as the installation of a single cell tower or the deployment of fiber optic cables, where the project site falls within the historical or cultural territory of several tribes. Each tribe may charge a separate fee for reviewing the project and issuing a clearance letter, even when the assessments are redundant and the likelihood of significant cultural or historical impact is minimal.

For instance, a telecommunications company seeking to install a single cell tower in a rural area might encounter clearance costs from several tribes, each charging fees that typically range from $2,000 to $10,000 per tribe. If the project site requires consultations with five or more tribes, the cumulative costs can quickly escalate to tens of thousands of dollars, even for a straightforward installation with negligible environmental or cultural risks (Smith & Johnson, 2020). This situation is further exacerbated when each tribe conducts its review independently, often duplicating research and administrative efforts that could potentially be streamlined or consolidated. I personally experienced Native American tribes charge in excess of over $20,000 per site as an aggregate cost across all of the tribes listed in area.

These overlapping fees are not limited to new installations. Modifications to existing infrastructure, such as upgrading antennas or replacing equipment, can trigger the same clearance requirements, with each involved tribe imposing fees despite the minimal changes to the site's footprint or impact. In some cases, tribes have charged fees repeatedly for projects involving minor adjustments to already-reviewed infrastructure, compounding costs over time.

The financial burden of multiple clearance fees not only impacts telecommunications companies but also slows the pace of infrastructure deployment, particularly for essential 5G networks that rely on dense installations of small cells. Delays caused by negotiating and paying clearance fees to multiple tribes can significantly extend project timelines, further increasing costs. Additionally, these expenses are typically passed on to consumers, resulting in higher service fees and phone bills.

While the consultation process is vital for preserving cultural and historical heritage, the lack of standardized procedures or fee structures creates inefficiencies and inequities in the system. Developing a centralized framework for managing tribal consultations or implementing reasonable, standardized fee caps could address these challenges, ensuring both cultural preservation and the efficient deployment of critical telecommunications infrastructure (FCC, 2023; Smith & Johnson, 2020).

Examples of Tribes Charging Fees Repeatedly for Minor Modifications

One of the most challenging aspects of the NEPA consultation process involving Native American tribes is the repeated imposition of fees for projects undergoing minor modifications. Telecommunications projects often require periodic updates to maintain or improve functionality, such as replacing outdated antennas, upgrading equipment, or adding new technologies to existing infrastructure. Despite these minor changes having negligible environmental or cultural impacts, they still trigger the same consultation requirements as new installations. This often leads to multiple tribes charging fees each time a modification is proposed, regardless of the scale or significance of the change (FCC, 2023).

For example, a telecommunications company seeking to upgrade antennas on an existing cell tower may need to consult with several tribes, each conducting a separate review and charging a fee for their assessment. These fees, which typically range from $2,000 to $10,000 per tribe, are imposed even if the site was previously reviewed and cleared during the initial installation. In some cases, the modifications do not alter the tower's footprint or affect surrounding land, making the need for additional cultural assessments redundant. Yet, because federal regulations mandate these consultations for projects tied to FCC licensing, companies have no choice but to comply (Smith & Johnson, 2020).

In another instance, routine maintenance on a tower in a previously assessed area required new clearance letters from several tribes. Each tribe charged fees for reviewing the project, even though the maintenance work involved replacing existing equipment with newer models and did not disturb the land or alter the site's structural integrity. This duplication of effort resulted in significant delays and inflated costs for what was essentially a low-impact activity (CEQ, 2020).

These repetitive charges can accumulate quickly, especially for large telecommunications networks that require ongoing upgrades across multiple sites. For example, deploying 5G technology, which involves densifying existing networks with small cells and upgrading older infrastructure, has been particularly affected by repeated tribal fees. A single company rolling out 5G in a region may face hundreds of thousands of dollars in cumulative fees, as each small cell or upgrade triggers new consultations and associated costs (FCC, 2023). This not only increases the financial burden on telecommunications providers but also delays the rollout of critical infrastructure.

The repetitive nature of these charges highlights inefficiencies in the current system. While it is essential to respect tribal sovereignty and ensure that culturally significant sites are protected, the lack of differentiation between major projects and minor modifications creates unnecessary hurdles. Developing streamlined processes, such as categorical exclusions for routine upgrades or a centralized system for managing consultations, could mitigate these issues. Standardized fee structures or one-time, site-based reviews could also help balance the need for cultural preservation with the practical realities of modern telecommunications infrastructure deployment. Without such reforms, the cycle of repetitive fees will continue to burden the industry and slow technological progress (Smith & Johnson, 2020).

The Financial Burden on Telecommunications Companies and Consumers

The financial implications of clearance fees for telecommunications companies are profound, with ripple effects that extend directly to consumers. Clearance fees charged by tribes, often ranging from $2,000 to $10,000 per consultation, accumulate rapidly for large-scale projects. When combined with the high costs of hiring environmental consulting firms to manage NEPA compliance, these expenses contribute significantly to project budgets. For telecommunications companies, which often operate on tight margins to remain competitive, these costs are invariably passed on to consumers through higher service fees and monthly phone bills (CEQ, 2020).

Consider a scenario in which a telecommunications provider seeks to deploy a network of 5G small cells across a metropolitan area. Each small cell requires a NEPA review and multiple tribal consultations, even if the cells are installed on pre-existing utility poles in urban areas. For a network requiring hundreds or thousands of installations, the cumulative costs of tribal fees and environmental consulting services can reach millions of dollars. These costs become part of the overall project budget, which telecommunications companies recover by increasing prices for services like data plans and mobile subscriptions (FCC, 2023).

The financial burden is exacerbated for smaller telecommunications companies that lack the resources of industry giants. These smaller providers are particularly vulnerable to high clearance and compliance costs, which can deter them from expanding their networks or force them to abandon planned upgrades. The resulting lack of competition in the market can lead to higher prices for consumers, compounding the financial strain caused by excessive regulatory fees (Smith & Johnson, 2020).

For consumers, the impact of these costs is tangible. Monthly phone bills, which already include charges for infrastructure maintenance and upgrades, are further inflated by the need to cover the expenses associated with NEPA compliance. Over time, these incremental increases add up, disproportionately affecting low-income households and those in underserved areas where infrastructure upgrades are most needed. Moreover, the delays caused by protracted NEPA reviews and consultations can slow the rollout of new technologies like 5G, depriving consumers of improved connectivity and service quality while they continue to pay higher prices (CEQ, 2020).

The financial burden extends beyond consumers to broader economic impacts. Delayed infrastructure deployment affects businesses reliant on advanced telecommunications services, such as high-speed internet and IoT technologies, which are critical for modern commerce and innovation. The inefficiencies and added costs tied to NEPA compliance create a drag on economic growth, particularly in regions where improved telecommunications infrastructure could drive development and investment.

Reforming the NEPA process to streamline consultations and establish reasonable, standardized fees for clearance letters could alleviate some of these financial pressures. By reducing unnecessary reviews for low-impact projects and ensuring that fees are commensurate with the scope and risk of the project, policymakers could help mitigate the economic burden on both companies and consumers. Without such changes, the cumulative costs of compliance will continue to hinder infrastructure development and inflate consumer expenses (FCC, 2023).

The Complexities of Tribal Involvement

The involvement of Native American tribes in the NEPA review process reflects a vital effort to preserve cultural heritage and ensure that federally regulated projects respect tribal lands and artifacts. However, this process has grown increasingly complex, introducing challenges for both telecommunications companies and tribal entities. While the consultation requirement under federal law aims to protect historically and culturally significant sites, the practical implementation often results in overlapping jurisdictions, inconsistent fee structures, and prolonged negotiations. These complexities not only delay critical infrastructure projects but also contribute to inflated costs, raising questions about how to balance legitimate cultural preservation with the efficient deployment of modern telecommunications infrastructure. Understanding the intricate dynamics of tribal consultations is essential for navigating the NEPA process while maintaining respect for tribal sovereignty and heritage.

Explanation of Why Tribal Input Is Required

Tribal input is an essential component of federally regulated projects, mandated to ensure the preservation of culturally significant sites and the protection of tribal heritage. This requirement stems from laws such as the National Historic Preservation Act (NHPA) and the National Environmental Policy Act (NEPA), both of which emphasize the need to safeguard historical and cultural resources during the planning and execution of federally licensed projects. For telecommunications infrastructure, particularly cell towers, tribal consultations are a critical step in ensuring that construction activities respect sacred sites, burial grounds, and areas of archaeological importance (FCC, 2023).

Cell towers, by their nature, can pose a risk to culturally sensitive areas due to their physical presence and the construction activities required for their installation. These structures can disrupt landscapes, impact historical landmarks, and infringe upon sacred or ceremonial spaces that hold deep significance for Native American tribes. For instance, a tower proposed near a sacred burial site may necessitate extensive tribal input to ensure that construction does not disturb human remains or ceremonial artifacts. Similarly, towers situated in regions with a rich tribal history may require archaeological assessments to identify and protect artifacts or other evidence of ancestral activities. The visual impact of towers can also raise concerns, particularly when they are located near culturally significant vistas or landmarks, where their presence could alter the spiritual or historical context of the area.

The consultation process ensures that tribes have a voice in decisions that may affect their cultural heritage, aligning with broader efforts to uphold tribal sovereignty and preserve the nation's diverse historical and cultural fabric. Federal agencies, such as the Federal Communications Commission (FCC), require project developers to engage with affected tribes, enabling them to provide input and recommendations for safeguarding sensitive sites. This consultation process reflects a recognition of the unique relationship between the federal government and Native American tribes, as well as the moral and legal obligations to protect tribal heritage.

However, the application of tribal consultation requirements has expanded significantly, creating complexities that go beyond the original intent of the law. While the need for tribal input is crucial in areas with known cultural or historical significance, it is now mandated even for projects in locations where the likelihood of affecting sensitive sites is minimal. For example, cell towers in urbanized or heavily developed areas often face the same consultation requirements as those in rural or undeveloped regions, despite the vastly different risks to cultural resources. These blanket requirements can lead to unnecessary delays and costs, diverting resources from projects that may genuinely threaten tribal heritage.

Moreover, the consultation process itself can become complicated due to overlapping tribal claims, the varying standards and fees imposed by different tribes, and the involvement of multiple federal and state agencies. Telecommunications companies often face challenges in navigating these overlapping jurisdictions and addressing the diverse requirements of the tribes involved. For projects that cross multiple tribal territories, developers may be required to conduct separate consultations and secure individual clearance letters from each tribe, significantly increasing administrative burdens and costs.

While the intent behind tribal consultation requirements is rooted in respect and preservation, the complexities of their application in modern infrastructure projects—particularly in the telecommunications sector—highlight the need for balance. Striking a balance between protecting cultural heritage and enabling the efficient deployment of critical infrastructure like cell towers is essential. This balance can ensure that the intent of tribal consultation is upheld while reducing unnecessary burdens for projects with minimal impact on culturally significant areas.

Discussion of Legitimate Cultural Preservation Concerns

While the preservation of cultural heritage is undoubtedly critical, the process of integrating tribal consultations into federally regulated projects must strike a balance between respecting legitimate concerns and ensuring practical, efficient implementation. Cultural preservation efforts are rooted in a recognition of the importance of safeguarding sacred sites, historical landmarks, and other culturally significant areas. However, when procedural requirements lead to excessive fees and prolonged delays, they can inadvertently erode the collaborative spirit that laws like NEPA and the National Historic Preservation Act (NHPA) were designed to foster.

Excessive fees imposed by some tribal groups for consultation and clearance letters can create financial burdens that extend far beyond what is necessary to cover administrative costs. While these fees are intended to compensate for the time and resources required to review projects, instances of exorbitant charges—sometimes reaching tens of thousands of dollars for a single project—have raised concerns about fairness and consistency. For telecommunications companies, these costs accumulate quickly, particularly for large-scale projects involving multiple towers or small cell installations that each require tribal clearance. In turn, these financial burdens are passed along to consumers, driving up the cost of phone bills and hindering the equitable deployment of critical infrastructure like 5G networks (FCC, 2023).

In addition to financial challenges, procedural delays can significantly impact project timelines, particularly when consultations require coordination with multiple tribes, federal agencies, and consulting firms. Prolonged reviews for projects with minimal environmental or cultural risks create bottlenecks that delay the deployment of telecommunications infrastructure. These delays can hinder advancements in connectivity, leaving underserved areas without access to high-speed internet or reliable cellular networks. Such outcomes are particularly problematic in rural and economically disadvantaged communities, where infrastructure upgrades are urgently needed to bridge the digital divide.

The risk of undermining NEPA's collaborative intent arises when the process is perceived as overly bureaucratic or exploitative. NEPA was designed to ensure that all stakeholders—government agencies, project developers, and affected communities, including Native American tribes—could engage in meaningful dialogue and work together to protect the environment and cultural heritage. When the process becomes mired in excessive costs or procedural inefficiencies, it risks fostering resentment rather than collaboration. This can lead to a breakdown in trust and a perception that the system prioritizes procedural compliance over substantive protection.

Balancing the need for cultural preservation with practical implementation requires a reevaluation of current processes to ensure fairness, efficiency, and mutual respect among all parties involved. Clearer guidelines for determining reasonable fees, streamlined procedures for low-impact projects, and increased federal oversight to address inconsistencies could help restore the collaborative intent of NEPA. By aligning cultural preservation goals with practical implementation, the process can remain true to its purpose while supporting the timely and cost-effective deployment of critical infrastructure.

How Consulting Firms and Mediators Complicate the Process

Consulting firms and third-party mediators often play a pivotal role in navigating the complexities of NEPA compliance, particularly in projects requiring tribal consultations. While their involvement is intended to streamline negotiations and ensure all parties meet regulatory requirements, these intermediaries can inadvertently (or deliberately) complicate the process, adding layers of financial and procedural burden to projects. This dynamic transforms what should be straightforward consultations into a lucrative business, driven in part by the self-interest of the industry.

In many cases, consulting firms are contracted to oversee the entire compliance process, from preparing Environmental Assessments (EAs) and Environmental Impact Statements (EISs) to facilitating communication between developers and Native American tribes. These firms often recommend third-party mediators to handle the delicate negotiations over tribal clearance letters, particularly when multiple tribes are involved or when disputes over fees arise. While these mediators may help resolve conflicts, they typically charge substantial fees for their services, which are added to already inflated project budgets. The involvement of multiple intermediaries not only increases costs but also creates additional layers of bureaucracy, further delaying project timelines (CEQ, 2020).

The financial incentives for consulting firms and mediators in this system are significant. With vague NEPA guidelines and the regulatory requirement for tribal input, these intermediaries operate in a landscape where their services are not just helpful but often essential. By leveraging the complexity of the regulatory framework, consulting firms can justify repeated or overlapping studies, even for low-impact projects. For example, a firm might suggest multiple assessments—such as separate cultural and environmental reviews—when a unified report would suffice. Similarly, mediators may prolong fee negotiations to increase their billable hours, compounding delays and inflating costs without necessarily providing substantive value to the project (Smith & Johnson, 2020).

Industry self-interest further drives this lucrative business model. Consulting firms often position themselves as indispensable experts, emphasizing the risks of non-compliance to developers. This approach, while partially valid, also capitalizes on developers' fear of litigation or project rejection, encouraging them to err on the side of over-compliance. In doing so, firms effectively perpetuate a system where every project—regardless of its environmental or cultural impact—

becomes an opportunity for additional revenue. Mediators, in turn, benefit from the prolonged and contentious nature of some tribal negotiations, with their fees increasing in proportion to the complexity and duration of the process.

The financial burden created by this dynamic is ultimately passed on to consumers. Telecommunications companies, facing mounting costs from consulting fees, mediator charges, and tribal clearance expenses, often recoup these losses by increasing service fees. This drives up the cost of phone bills, effectively transferring the financial impact of a convoluted regulatory process to the end user. Moreover, the delays caused by prolonged consultations and redundant reviews hinder the timely deployment of critical infrastructure, particularly for 5G networks and other advanced telecommunications technologies.

Reforming this system requires addressing the self-interested incentives that drive excessive costs. Clearer NEPA guidelines, stricter oversight of consulting practices, and standardized tribal fee structures could help reduce the reliance on costly intermediaries. Additionally, fostering direct communication between developers and tribes—without unnecessary third-party involvement—could streamline the process and rebuild trust among stakeholders. By minimizing the role of self-interest in NEPA compliance, the system could become more efficient, equitable, and focused on its core mission of protecting the environment and cultural heritage.

Impact on Consumers Due to Project Delays and Inflated Compliance Costs

The intricate and often redundant processes associated with NEPA compliance in the telecommunications sector create a ripple effect that ultimately impacts consumers. Delays caused by lengthy environmental reviews, tribal consultations, and excessive fees not only slow down the deployment of critical infrastructure like 5G networks but also inflate project budgets. These costs, absorbed by telecommunications companies during project development, are inevitably passed on to consumers in the form of higher service fees and phone bills. The result is a system where regulatory inefficiencies and inflated compliance costs hinder technological progress and burden end users financially, raising questions about the balance between environmental protection, cultural preservation, and the practical needs of a connected society.

How Excessive Fees and Consulting Services Drive Up Costs

The financial impact of excessive fees and consulting services in the NEPA compliance process directly affects consumers by significantly driving up the costs of telecommunications projects. Telecommunications companies, faced with rising expenses due to clearance fees, consulting firm charges, and prolonged project timelines, often have no choice but to pass these costs on to their customers. This results in higher monthly phone bills and service fees, disproportionately impacting low-income households and those in rural areas where infrastructure development is already more expensive (FCC, 2023).

For instance, the cumulative costs of tribal clearance letters, which can reach tens of thousands of dollars for a single project, and additional consulting fees for redundant environmental reviews inflate the overall budget for installing or upgrading cell towers. These increased costs ripple through the telecommunications sector, ultimately making services less affordable for millions of Americans. The issue is particularly concerning as 5G technology and other advancements in connectivity become essential for everyday life, widening the digital divide between those who can afford rising service fees and those who cannot.

Moreover, the inefficiencies in the NEPA process exacerbate the financial burden. Prolonged delays mean that companies face extended timelines for recouping their investments, further incentivizing them to adjust consumer prices. The combination of higher costs and slower project implementation undermines the industry's ability to expand access to high-speed wireless networks, which are critical for economic growth and equitable connectivity. Without reforms to streamline compliance for low-risk projects, the financial strain on consumers is likely to persist, creating barriers to widespread telecommunications access.

Delays Caused by Lengthy Negotiations

Negotiations over clearance fees and the prolonged review process significantly hinder the timely deployment of critical telecommunications infrastructure, with particularly acute effects on the rollout of 5G networks. These delays, often caused by protracted discussions between developers, tribes, and consulting firms, disrupt project timelines and stall the implementation of advanced technologies. For 5G networks, which require a dense network of small cell installations to deliver high-speed, low-latency connections, any slowdown in deployment can have far-reaching consequences for connectivity and technological innovation (FCC, 2023).

The impact of these delays extends beyond the telecommunications industry, affecting industries reliant on 5G technology, such as healthcare, transportation, and manufacturing. For instance, the adoption of autonomous vehicles, smart city infrastructure, and remote healthcare services depends heavily on the availability of reliable 5G networks. Every delay in infrastructure deployment pushes back the timelines for these advancements, slowing progress in sectors poised to drive economic growth and improve quality of life (Smith & Johnson, 2020).

Additionally, the financial implications of these delays compound the problem. Telecommunications companies incur higher project costs as they navigate drawn-out review processes, which often include renegotiating clearance fees, repeating environmental assessments, and addressing administrative hurdles. These increased costs, coupled with the extended time needed to complete projects, result in higher operational expenses. To offset these financial burdens, companies frequently pass the costs on to consumers in the form of increased service fees, further inflating phone bills and reducing the affordability of telecommunications services for everyday users (CEQ, 2020).

The prolonged review process also creates competitive disadvantages on a global scale. Countries with more streamlined regulatory frameworks for telecommunications infrastructure are able to advance 5G networks and related innovations more quickly, leaving the U.S. at risk of falling behind in the race to develop cutting-edge technologies. Without reforms to expedite negotiations and streamline compliance processes, the inefficiencies inherent in the current system will continue to hinder the nation's ability to maintain leadership in the global digital economy. Addressing these delays is essential not only for reducing costs but also for fostering the technological advancements that underpin modern connectivity and innovation.

The Ripple Effect on Consumers

The compounding costs of compliance in the telecommunications industry create a significant financial burden that ultimately falls on end-users, affecting the affordability and accessibility of vital services. Every layer of regulatory compliance—including tribal clearance fees, consulting firm expenses, and delays stemming from prolonged NEPA reviews—adds to the overall cost of deploying new infrastructure. These increased expenses ripple through the industry, as telecommunications providers incorporate these costs into their operational budgets, which are then passed along to consumers in the form of higher monthly phone bills and service fees.

For instance, the fees associated with securing clearance letters from multiple tribes, often ranging from thousands to tens of thousands of dollars per project, accumulate rapidly when multiplied across the thousands of cell towers required for modern networks. Similarly, consulting firms, which charge premium rates for environmental assessments and documentation, drive up project costs even for minor infrastructure upgrades. These financial pressures are exacerbated by delays caused by negotiations, redundant reviews, and administrative hurdles, which extend project timelines and inflate labor and overhead expenses (FCC, 2023).

The cumulative effect of these costs disproportionately impacts low-income and rural communities, where the expansion of telecommunications infrastructure is already financially challenging. Providers may deprioritize deploying services to these areas, deeming the costs too prohibitive to justify the investment. This dynamic deepens the digital divide, leaving underserved populations without access to affordable, reliable telecommunications services and hindering their ability to participate fully in the digital economy.

Efforts to expand affordable telecommunications nationwide are further undermined by the industry's need to recoup compliance-related expenses. Initiatives aimed at reducing service costs or offering subsidized access are compromised when companies must allocate resources to cover regulatory expenditures. These financial burdens also divert funds that could be invested in research, development, and network enhancements, slowing innovation and reducing the industry's ability to improve service quality and expand coverage efficiently (CEQ, 2020).

Moreover, the financial impact is not limited to individual consumers. Businesses and organizations that rely on robust telecommunications networks also face higher costs, particularly in industries like healthcare, education, and transportation that depend on advanced connectivity solutions. These increased expenses are passed on through higher operational costs, affecting prices for goods and services across various sectors.

Ultimately, the compounding costs of compliance create a cascading effect that undermines the broader goal of expanding affordable telecommunications services nationwide. Reforming regulatory frameworks to streamline compliance, reduce unnecessary fees, and prioritize efficiency is essential to mitigate these financial impacts and ensure that modern telecommunications infrastructure can be developed equitably and sustainably.

Case Studies Illustrating Egregious Overreach and Inflated Project Budgets

The application of NEPA to the telecommunications industry has generated numerous examples of regulatory overreach, where routine projects incur excessive costs and delays due to unnecessary environmental reviews. These case studies highlight how NEPA, originally intended to safeguard significant environmental resources, has been applied in ways that add little environmental value but significantly inflate project budgets. From redundant assessments for minor upgrades to inflated tribal clearance fees, these examples expose the inefficiencies embedded in the current compliance process. Such overreach not only strains telecommunications companies but also hinders the timely deployment of critical infrastructure, such as 5G networks, ultimately burdening consumers with higher costs and slower service expansion. By examining these cases, this chapter underscores the urgent need for reforms to ensure that NEPA compliance aligns with the scope and impact of the projects it regulates.

Example 1: Urban Tower Projects

In one striking example of regulatory overreach, a proposed cell tower in a densely populated urban area, surrounded by pre-existing infrastructure such as utility poles and streetlights, was subjected to multiple cultural and environmental reviews. Despite its negligible environmental impact and minimal risk of disturbing culturally significant sites, the project faced extensive compliance requirements. Federal regulations mandated clearance letters from several Native American tribes, each of which charged separate fees for their assessments. The total cost for these letters alone amounted to $25,000, with additional administrative expenses incurred through consulting firms that facilitated the process. These firms often serve as intermediaries, guiding compliance while adding to project budgets (FCC, 2023).

The procedural requirements also extended the project timeline significantly, delaying construction by six months. During this period, the telecommunications company faced mounting operational costs, including fees for ongoing regulatory consultations and project management. Beyond financial burdens, the delay disrupted plans to improve network coverage in the urban area, leaving residents and businesses without much-needed enhancements to connectivity and bandwidth. This example illustrates how the misapplication of the National Environmental Policy Act (NEPA) to low-impact projects can lead to substantial costs and inefficiencies, even in scenarios where environmental and cultural risks are minimal.

Furthermore, the lack of standardized fees or clear procedural limits exacerbates these issues. In this case, multiple tribes assessed the same project independently, and there was no mechanism to consolidate or coordinate their reviews, leading to redundant and costly assessments. These inefficiencies highlight a systemic problem within NEPA's implementation for federally licensed projects, where overlapping reviews and unclear thresholds for significant impact inflate costs without offering corresponding environmental or cultural benefits (Smith & Johnson, 2020).

Ultimately, the financial burden from these excessive fees and delays does not remain confined to the telecommunications providers. Instead, these costs are passed on to consumers in the form of higher service fees and phone bills, undermining efforts to make telecommunications services more affordable and accessible to all Americans. This case emphasizes the urgent need for regulatory reform to streamline NEPA compliance processes for low-impact projects, ensuring that environmental and cultural protections are balanced against practical and financial realities.

Even minor modifications to existing telecommunications infrastructure, such as an antenna upgrade on a pre-existing tower, can trigger new NEPA reviews and compliance requirements. In one case, a telecommunications company seeking to replace aging antennas with more efficient equipment encountered a host of regulatory hurdles. Although the project did not alter the tower's footprint, height, or operational characteristics, NEPA compliance was still mandated due to its federal licensing through the Federal Communications Commission (FCC).

As part of the review, the company was required to conduct environmental assessments, cultural reviews, and consultations with multiple Native American tribes. Each tribe assessed whether the project could potentially impact culturally significant sites, despite the minimal changes involved. Fees for tribal clearance letters alone reached $30,000, with additional costs for consulting firms managing the compliance process. The total project compliance costs exceeded $50,000, significantly inflating the budget for what would otherwise have been a routine maintenance upgrade. These fees were further compounded by the time delays caused by redundant reviews and protracted negotiations over clearance requirements, which extended the project timeline by several months (FCC, 2023; Smith & Johnson, 2020).

This example underscores a broader issue with NEPA's application to low-impact projects. While the original intent of NEPA was to address significant environmental risks, its extension to minor upgrades creates inefficiencies that burden the telecommunications sector without yielding substantive environmental or cultural benefits. The project's compliance requirements, which were disproportionate to its environmental impact, illustrate how vague regulatory guidelines enable the consulting and clearance process to escalate in both scope and cost.

Furthermore, the financial burden of these compliance costs is ultimately borne by consumers. Telecommunications providers pass these expenses onto users through higher service fees, contributing to the rising cost of cellular services. This dynamic not only affects the affordability of telecommunications but also hampers efforts to modernize infrastructure quickly, particularly for 5G deployment, which relies on incremental upgrades to existing networks. Reforming NEPA's application to minor projects could streamline the process, reduce costs, and maintain environmental and cultural protections while encouraging infrastructure development that benefits consumers and communities.

Example 2: Clearwire Project Buildout in the Early 2010s

The Clearwire project, a nationwide initiative to expand broadband internet services in the early 2010s, encountered significant regulatory obstacles due to the misapplication of environmental and cultural review requirements. Despite its mission to provide high-speed internet access to underserved areas, many of its proposed infrastructure deployments faced costly delays and inflated expenses, largely driven by outdated regulatory frameworks like the National Environmental Policy Act (NEPA).

Clearwire aimed to deploy approximately 16,000 towers and antenna systems nationwide, with installations ranging from small cell sites to larger antennas on existing structures such as utility poles, rooftops, and pre-established towers (FCC, 2012). These projects typically had minimal environmental impact, as they did not require significant ground disturbance or alterations to local ecosystems. However, because they were federally licensed through the Federal Communications Commission (FCC), they were subject to the same NEPA compliance requirements as large-scale infrastructure projects like highways or power plants. This blanket application of NEPA regulations created unnecessary barriers to Clearwire's efforts to modernize broadband infrastructure.

One Clearwire project in the Midwest highlighted these challenges. A proposed antenna installation on an existing 80-foot tower triggered a series of environmental and cultural assessments, despite the fact that the project would not alter the tower's footprint or height. Compliance requirements included consultations with multiple Native American tribes to assess potential impacts on culturally significant sites. These consultations came with hefty fees, as each tribe conducted independent reviews, charging a combined $35,000 for clearance letters. Additionally, Clearwire had to hire environmental consulting firms to manage the compliance process, incurring another $25,000 in administrative costs.

The project also faced significant delays. The overlapping and redundant reviews extended the timeline by more than eight months, during which Clearwire incurred operational expenses, including leasing fees for the proposed site and ongoing consulting services. These delays also affected the broader community, as the project was designed to improve broadband access in rural and underserved areas. Residents and businesses were forced to wait nearly a year for upgraded internet services, impacting education, healthcare access, and economic development in the region.

The Clearwire example underscores the inefficiencies of applying NEPA to low-impact telecommunications projects. While NEPA was originally designed to address substantial environmental risks, its use for routine infrastructure projects inflates costs without delivering proportional environmental or cultural benefits. The cumulative cost of compliance for the Clearwire project exceeded $75,000—an expense that was ultimately passed on to consumers through higher broadband service fees.

Clearwire's nationwide deployment faced similar challenges. With tens of thousands of sites requiring redundant reviews and compliance documentation, the cumulative costs of NEPA compliance significantly escalated the overall expense of the project. These unnecessary costs contributed to higher operational expenses for telecommunications providers and, ultimately, more expensive internet services for consumers (Smith & Johnson, 2014).

This case also highlights the structural flaws in the current regulatory framework. Clearwire's projects faced duplicative reviews from multiple agencies and tribes, with no streamlined mechanism to consolidate or coordinate these processes. The lack of clear thresholds for what constitutes a "significant impact" under NEPA enables these inefficiencies to persist, disproportionately burdening small and mid-sized telecommunications projects.

Reforming NEPA's application to projects like Clearwire's could yield significant benefits for both the industry and consumers. By exempting low-impact installations from the full scope of NEPA compliance, regulatory agencies could reduce costs, expedite project timelines, and encourage infrastructure development that directly benefits underserved communities. At a time when access to high-speed internet is essential for education, work, and healthcare, applying a 50-year-old regulatory policy to modern technologies is not just outdated—it is irresponsible. Modernizing these regulations would balance the need for environmental and cultural protections with the practical realities of today's telecommunications landscape, ensuring that progress is not stalled by inefficiencies.

Example 3: Disparate Fee Structures

Clearance fees charged by Native American tribes for telecommunications projects exhibit significant variability, even for projects with similar scopes and minimal environmental or cultural impacts. For instance, while one tribe may charge a relatively modest $2,000 for a clearance letter, another might demand upwards of $10,000 for an identical project. This inconsistency in fee structures creates uncertainty for telecommunications companies, complicating project budgeting and timelines.

The lack of standardized fee guidelines allows for significant disparities, often unrelated to the actual effort or resources required to evaluate a project's potential impact on culturally significant sites. Some tribes justify higher fees by citing administrative costs, while others may leverage their jurisdictional authority to maximize revenue. While it is essential to honor and respect tribal sovereignty and the need for cultural preservation, this unregulated variation in fees frequently leads to financial strain on project developers, particularly for those operating in multiple jurisdictions.

This inconsistency underscores the urgent need for standardized fee structures that balance fair compensation for tribes with the practicalities of infrastructure development. Clear and reasonable fee guidelines could streamline the consultation process, reduce disputes, and foster better collaboration between tribes and the telecommunications industry. Furthermore, a more predictable cost framework would allow for efficient project planning, ultimately benefiting consumers by minimizing unnecessary expenses passed down through service fees.

Proposed Solutions: Addressing NEPA Inefficiencies in the Telecommunications Sector

The challenges posed by NEPA's application to telecommunications projects demand thoughtful and balanced reforms. Excessive fees, redundant reviews, and prolonged delays have highlighted inefficiencies that burden the industry and ultimately impact consumers. Proposed solutions aim to streamline the NEPA compliance process, reduce unnecessary costs, and foster collaboration between stakeholders. By implementing standardized fees, categorical exclusions for low-impact projects, and transparent consultation frameworks, policymakers can ensure that NEPA continues to protect cultural and environmental resources without stifling critical infrastructure development. These reforms not only address systemic inefficiencies but also pave the way for affordable and reliable telecommunications services to meet the growing demands of the digital age.

Standardized Fees for Tribal Clearance Letters

Establishing standardized fees for tribal clearance letters is an essential reform to address the financial disparities that currently plague the NEPA compliance process. At present, the fees charged by tribes for reviewing telecommunications projects vary widely, ranging from as little as $2,000 to more than $20,000 for comparable projects (Smith & Johnson, 2020). This inconsistency not only creates uncertainty for project developers but also inflates budgets, making it challenging for smaller telecommunications companies to compete in the market. Additionally, these unpredictable costs ultimately trickle down to consumers, contributing to higher service fees and phone bills.

A standardized fee structure would provide much-needed clarity and equity in the process, ensuring that tribes are fairly compensated for their time and expertise while preventing exorbitant charges that do not align with the scope or impact of the projects. Such a system would reduce instances where multiple tribes charge overlapping fees for the same project, particularly in cases where cultural or historical significance is minimal. This reform would also promote consistency across the industry, helping to eliminate the financial uncertainties that currently delay project timelines and escalate costs.

Implementing standardized fees would require collaboration between federal agencies, such as the Federal Communications Commission (FCC), and tribal governments to develop transparent guidelines that respect tribal sovereignty while addressing industry concerns. This process should involve public consultations to ensure that all stakeholders—tribes, developers, and consumers—have input on the fee structures. A tiered system could also be considered, where fees are scaled based on the complexity and scope of the project, ensuring that smaller, low-impact installations do not incur the same costs as large-scale developments.

In addition to addressing cost disparities, standardized fees would help reduce disputes and delays often associated with negotiating clearance charges. The current system, where developers are sometimes forced into prolonged negotiations to resolve discrepancies, diverts focus from the primary goal of cultural preservation. By providing clear, predictable costs upfront, a standardized approach would allow both developers and tribes to streamline their processes, ensuring that resources are allocated efficiently.

Finally, these reforms could be integrated into a centralized consultation system, overseen by a neutral federal body, to further enhance transparency and accountability. Such a system would allow developers to access information about standard fees, monitor the progress of consultations, and ensure that payments are directed appropriately. This streamlined approach would strengthen trust among stakeholders, reduce administrative burdens, and reinforce the importance of meaningful cultural preservation over financial contention.

Categorical Exclusions for Urban or Low-Impact Projects

Categorical exclusions (CEs) for urban or low-impact projects present a logical and efficient solution to the problem of redundant NEPA reviews for installations with minimal environmental or cultural risks. Under NEPA, CEs are provisions that exempt certain types of projects from extensive environmental reviews when it is clear that their impact will be negligible (CEQ, 2020). Extending the application of CEs to telecommunications projects in already developed areas—such as urban neighborhoods, industrial zones, or pre-existing utility infrastructure—would significantly reduce procedural bottlenecks while maintaining NEPA's overarching goals of environmental protection and public accountability.

The telecommunications industry, particularly with the rapid deployment of 5G networks, stands to benefit greatly from such reforms. Small cell installations and minor upgrades to existing towers, for example, often involve minimal or no changes to the physical footprint of a site. However, under current guidelines, these projects frequently undergo the same rigorous review process as new tower installations in environmentally or culturally sensitive areas. This one-size-fits-all approach delays projects, increases costs, and diverts resources away from more impactful environmental concerns. By categorizing low-risk projects as eligible for CEs, policymakers could expedite critical infrastructure deployment while ensuring that NEPA compliance focuses on projects with genuine potential for harm.

CEs for telecommunications projects would be particularly effective in urban areas, where existing development has already altered the landscape, reducing the likelihood of significant environmental or cultural impacts. For instance, small cell installations on utility poles in a downtown business district or upgrades to antennas on an existing tower in a commercial zone would be prime candidates for categorical exclusions. These projects are vital for improving network capacity and reliability, yet they often face procedural delays that hinder timely deployment. With CEs in place, these projects could move forward without the need for costly and time-consuming Environmental Assessments (EAs) or Environmental Impact Statements (EISs).

To ensure accountability and safeguard NEPA's objectives, policymakers could establish clear criteria for projects eligible for CEs. For example, CEs could be limited to projects in areas outside of culturally significant or ecologically sensitive zones, such as wetlands, historic landmarks, or tribal lands. This targeted approach would maintain the integrity of NEPA's mission to protect the environment and cultural heritage while reducing unnecessary administrative burdens on low-risk developments. Furthermore, policymakers could require project proponents to submit basic documentation confirming that the proposed activity meets CE criteria, providing transparency without the need for a full-scale review.

Implementing CEs for low-impact telecommunications projects would not only reduce delays but also alleviate financial burdens on both developers and consumers. The streamlined process would allow companies to allocate resources more efficiently, accelerating the rollout of advanced technologies like 5G while avoiding inflated costs that are often passed down to end-users. Additionally, by focusing NEPA resources on projects with significant environmental or cultural implications, agencies could enhance their capacity to conduct thorough reviews where they are truly needed.

The inclusion of CEs for urban and low-impact projects in NEPA compliance represents a balanced approach to modernizing the law in a way that reflects current technological and infrastructural realities. By adopting this reform, policymakers could facilitate infrastructure growth, promote innovation, and uphold environmental and cultural protections without compromising efficiency or fairness.

Centralized, Transparent Consultation System

A centralized system for managing tribal consultations would revolutionize the NEPA compliance process, offering a streamlined approach to reduce administrative burdens and improve transparency. Under the current framework, developers must individually engage with each tribe for clearance letters, a process that often results in inconsistent practices, redundant efforts, and significant delays. These inefficiencies can inflate project costs and create confusion for developers, particularly when projects span multiple tribal jurisdictions or require consultations with multiple tribes.

A centralized platform, overseen by a neutral federal entity such as the Federal Communications Commission (FCC), could serve as a comprehensive hub for managing all aspects of tribal consultations. This system could allow tribes to upload standardized fee schedules, clearly outline their review requirements, and maintain a real-time project tracking system. Developers, on the other hand, would gain access to a transparent and predictable process, with clear expectations for costs, timelines, and requirements. Such a system would eliminate much of the ambiguity that currently plagues tribal consultations, fostering a more efficient and equitable compliance environment.

Transparency would be a cornerstone of this centralized system. By requiring tribes to disclose standardized fees and provide clear criteria for review, the system could address growing concerns over excessive or arbitrary charges. Developers could compare costs across jurisdictions, ensuring fair pricing and preventing unnecessary financial strain. Similarly, tribes could benefit from greater visibility into the projects requiring their input, enabling them to allocate resources more effectively and maintain cultural preservation standards.

Additionally, a centralized platform could facilitate communication and collaboration between tribes, developers, and regulatory agencies. For example, the system could include automated notifications to streamline project updates, reduce redundant communications, and provide a digital record of interactions and decisions. This would help prevent delays caused by miscommunication or incomplete documentation, a frequent challenge under the current decentralized approach. Moreover, the centralized repository could serve as an archive for past consultations, reducing the need for repetitive reviews of previously assessed sites, such as those involving minor modifications to existing towers.

A centralized consultation system would also promote fairness and equity by ensuring that all stakeholders operate on a level playing field. Developers would no longer be at the mercy of inconsistent practices or unclear timelines, while tribes could rely on standardized processes to protect their cultural resources without unnecessary bureaucratic hurdles. Regulatory agencies, such as the FCC, would gain oversight capabilities to monitor consultations and intervene when disputes arise, ensuring compliance costs remain reasonable and aligned with NEPA's intent.

Ultimately, the creation of a centralized system for tribal consultations would reflect a modernized approach to NEPA compliance, addressing the inefficiencies that currently hinder telecommunications projects. By balancing the need for cultural preservation with the demands of infrastructure development, such a system would enhance collaboration, reduce delays, and lower costs for all parties involved, fostering a more efficient and sustainable path forward for the telecommunications industry.

FCC Oversight and Mediation

Enhanced oversight by the Federal Communications Commission (FCC) could significantly mitigate the challenges associated with exorbitant fees and disputes in the tribal consultation process. As the primary regulatory body for telecommunications projects, the FCC holds a unique position to enforce fairness and efficiency in compliance with the National Environmental Policy Act (NEPA). By taking a more active role in mediating conflicts and standardizing practices, the FCC could address many of the inefficiencies and inconsistencies that currently burden developers and tribes alike.

A formal dispute resolution mechanism established within the FCC could serve as an impartial platform for resolving disagreements over fees or consultation requirements. Such a mechanism would provide a clear avenue for developers and tribes to address conflicts without resorting to costly third-party mediators, which often inflate project budgets and prolong timelines. This process could involve mediation by FCC-appointed experts who ensure that disputes are resolved transparently and equitably, in line with federal regulations and NEPA's intent (Smith & Johnson, 2020).

In addition to resolving disputes, the FCC could take proactive steps by issuing clearer guidelines on what constitutes reasonable fees for tribal consultations. By defining acceptable cost ranges based on factors such as project scope, location, and cultural significance, the FCC could create a regulatory framework that discourages exploitative practices. These guidelines would provide much-needed clarity for both developers and tribes, reducing misunderstandings and fostering trust in the process. Developers would benefit from predictable and transparent fee structures, while tribes could confidently charge fees that reflect the actual costs of conducting reviews without fear of criticism or dispute.

Regular audits of consultation processes and fee structures could further enhance accountability. The FCC could review project records to identify patterns of inefficiency, redundancy, or excessive fees, allowing the agency to intervene when necessary. For example, audits could uncover instances where multiple tribes charge overlapping fees for the same project or where consultations are required for minor modifications with negligible impact. Such insights would enable the FCC to refine its oversight strategies and ensure that compliance processes align with the principles of efficiency and fairness.

Enhanced FCC oversight would also promote timely completion of projects, a critical factor for the telecommunications industry as it expands infrastructure for 5G networks and beyond. By addressing disputes promptly and ensuring that compliance costs remain reasonable, the FCC could minimize delays that currently hinder the deployment of critical telecommunications infrastructure. This approach would not only reduce costs for developers but also benefit consumers, who often bear the financial burden of inflated project expenses through higher service fees (FCC, 2023).

Ultimately, a more active FCC role in overseeing NEPA compliance would strike a balance between protecting tribal and environmental considerations and fostering the rapid, cost-effective expansion of telecommunications networks. By mediating disputes, issuing clear guidelines, and conducting audits, the FCC could enhance the efficiency and transparency of the compliance process, ensuring that the interests of all stakeholders are fairly represented.

In conclusion, while NEPA remains a cornerstone of environmental protection, its application in the telecommunications sector has increasingly become a mechanism for generating work and revenue for private consulting firms, Native American tribes, and regulatory oversight entities—all at the expense of the consumer. The very processes designed to safeguard cultural and environmental integrity have, in some cases, been exploited to inflate costs and create procedural bottlenecks, diverting resources from meaningful protections. These inefficiencies ripple through the industry, burdening telecommunications companies with inflated project budgets and delaying critical infrastructure projects like 5G deployment. Ultimately, it is the consumer who bears the brunt of these costs, reflected in higher service fees and phone bills. Addressing these systemic issues requires a balanced approach—one that preserves the integrity of NEPA's mission while introducing reforms to reduce unnecessary costs, enhance efficiency, and prioritize substantive environmental and cultural concerns over procedural redundancies.

Chapter 6

The Hidden Costs to Consumers

The hidden costs of NEPA compliance in the telecommunications sector are quietly impacting consumers, driving up phone bills and limiting access to affordable, reliable service. While designed to protect the environment, NEPA's extensive regulatory requirements—ranging from environmental reviews to tribal consultations—have introduced significant financial and operational burdens for telecommunications companies. These costs, compounded by permitting delays and prolonged project timelines, are ultimately passed on to consumers through higher service fees and carrier charges. Despite their invisibility on monthly bills, these expenses significantly influence the affordability of telecommunications services, raising important questions about the balance between regulatory oversight and consumer impact.

Permitting Delays and Compliance Costs

The regulatory requirements imposed by NEPA, particularly Environmental Assessments (EAs) and Environmental Impact Statements (EISs), impose significant and often underestimated financial burdens on telecommunications companies. These costs are multifaceted, encompassing direct expenses such as hiring environmental consulting firms, paying for tribal clearance letters, and completing detailed environmental documentation, as well as indirect costs related to project delays. Prolonged timelines for NEPA compliance lead to logistical challenges, including the need for extended equipment storage, prolonged workforce retention, and additional legal and administrative overhead. For large-scale infrastructure projects, delays of even a few months can translate into millions of dollars in unforeseen expenses. These cumulative costs are inevitably passed on to consumers through higher service fees and increased phone bills, further eroding the affordability of telecommunications services (FCC, 2023).

A significant contributor to these financial burdens is NEPA's indiscriminate application to both large-scale and low-impact projects. For instance, small cell installations—critical to the widespread adoption of 5G networks—are often treated with the same regulatory rigor as new, large-scale tower installations. This occurs despite the fact that small cells are frequently placed on pre-existing structures such as utility poles, streetlights, or building rooftops, with minimal environmental or cultural risks. Nonetheless, these projects often undergo the same exhaustive NEPA reviews as new builds in sensitive areas, including stakeholder consultations, tribal clearances, and public comment periods. This process not only inflates the project budget but also creates unnecessary bottlenecks, delaying infrastructure deployment and driving up costs with little to no added environmental benefit (Smith & Johnson, 2020).

The financial impact of these redundant and protracted reviews extends beyond the direct costs of compliance. Delayed deployment of critical infrastructure, such as 5G networks, introduces additional inefficiencies that further inflate expenses. For example, as project schedules extend, telecommunications companies incur higher storage costs for unused equipment, increased leasing fees for pre-approved sites, and prolonged workforce expenses, such as salaries and benefits for idle employees. Moreover, companies often face legal and administrative challenges during these delays, further driving up operational expenses. These rising costs are systematically transferred to consumers, effectively embedding NEPA-related inefficiencies into their monthly phone bills (CEQ, 2020).

Compounding these challenges is the evolving nature of telecommunications infrastructure, which demands faster deployment cycles to keep pace with technological advancements. Small cells and distributed antenna systems (DAS), essential for 5G networks, require dense deployments across urban and suburban areas. While these systems are relatively low-impact and environmentally benign, they frequently undergo extensive NEPA reviews. For example, minor equipment upgrades, such as replacing antennas on an existing tower, can trigger full environmental reviews, including redundant cultural and ecological assessments. These unnecessary processes inflate budgets, delay service improvements, and hinder the speed at which companies can meet growing consumer demand for faster and more reliable connectivity. As a result, consumers are left paying higher service fees for networks that lag behind their potential deployment pace (FCC, 2023).

Ultimately, NEPA compliance processes, as they are currently applied, create a cascading financial burden on the telecommunications industry and its customers. By subjecting even minor, low-risk projects to exhaustive reviews, NEPA's implementation results in inflated project budgets that are passed on to consumers through elevated carrier fees. These costs disproportionately affect low-income households, making access to affordable, high-quality telecommunications services increasingly unattainable. Reforming NEPA's approach to distinguish between high-impact and low-impact projects could mitigate these inefficiencies, ensuring that environmental protection remains a priority without unnecessarily burdening consumers. As the telecommunications sector continues to expand, addressing these challenges is critical to balancing environmental responsibility with the need for affordable, accessible services nationwide.

Increased Project Timelines Leading to Slowed Infrastructure Deployment

NEPA compliance requirements frequently result in substantial project delays, posing significant challenges for telecommunications companies tasked with building or upgrading critical infrastructure. These delays are particularly detrimental in the rollout of 5G networks, which depend on the deployment of thousands of small-cell installations to deliver the ultra-fast speeds and low latency required by modern connectivity demands. Unlike traditional cell towers, 5G small cells must be densely distributed across urban and suburban areas, often mounted on existing infrastructure such as utility poles or building rooftops. Despite their minimal environmental and cultural impact, these installations are often subjected to the same NEPA review processes as large-scale projects, including Environmental Assessments (EAs) and public comment periods. The procedural delays not only slow infrastructure deployment but also create a ripple effect across the industry, hindering technological advancement and innovation (CEQ, 2020).

One of the most significant consequences of these delays is the financial gap created by postponed revenue generation. For telecommunications companies, the ability to monetize new infrastructure depends on timely project completion. Delays caused by extended NEPA reviews mean that companies are unable to launch new services or expand coverage areas on schedule, resulting in lost opportunities to attract new customers or offer enhanced features to existing ones. To offset these financial losses, companies often turn to their existing customer base, passing on the costs through higher service fees or surcharges. This practice disproportionately affects consumers, who end up paying more for outdated or suboptimal services while waiting for the full benefits of advancements like 5G to materialize (Smith & Johnson, 2020).

The slow pace of infrastructure deployment due to NEPA-induced delays also stifles technological innovation. Technologies dependent on 5G, such as the Internet of Things (IoT), autonomous vehicles, and smart city applications, require a robust and expansive network to function effectively. Prolonged timelines for 5G deployment mean that these innovations are delayed, leaving consumers and industries unable to capitalize on their potential. For example, applications like telemedicine, which rely on high-speed, low-latency connections, are hindered in areas where 5G infrastructure has not yet been rolled out. The lack of timely access to cutting-edge technologies not only impacts individual users but also undermines economic growth and competitiveness in sectors reliant on advanced connectivity (FCC, 2023).

Additionally, the inefficiencies created by NEPA compliance affect the broader consumer experience. Telecommunications companies, burdened by regulatory costs and delays, often prioritize profitable urban markets while neglecting rural and underserved areas. This uneven deployment exacerbates the digital divide, leaving many communities without access to reliable high-speed internet. Consumers in these areas face higher costs for inferior services, as the delays associated with NEPA compliance make it less economically viable for companies to invest in low-density regions (CEQ, 2020).

Reforming NEPA to account for the unique characteristics of low-impact projects, such as small-cell installations, could mitigate these delays and enable faster infrastructure deployment. For instance, establishing categorical exclusions for projects in urban or suburban areas with minimal environmental risk would streamline the compliance process and allow telecommunications companies to complete projects more efficiently. Such reforms would benefit consumers by accelerating the rollout of advanced technologies, reducing service costs, and ensuring more equitable access to high-quality telecommunications services across all regions. Addressing these challenges is essential for balancing environmental stewardship with the need for innovation and affordability in the telecommunications sector.

Quantifying the Financial Burden on Consumers

The cumulative financial impact of NEPA compliance on the telecommunications sector is staggering, encompassing consulting fees, tribal clearance costs, and expenses related to project delays. Together, these factors drive up project budgets by billions of dollars annually, creating significant economic ripple effects throughout the industry. For instance, consulting firms engaged to navigate complex NEPA requirements charge substantial fees for Environmental Assessments (EAs) and Environmental Impact Statements (EISs), even for low-risk projects. Similarly, tribes often charge high fees for cultural clearance letters, with costs varying widely and sometimes exceeding $10,000 per project (Smith & Johnson, 2020). In addition, prolonged project timelines caused by NEPA compliance introduce further costs, including increased administrative overhead, equipment storage fees, and extended workforce management needs.

A recent study highlights that compliance-related expenses can account for 10–15% of total project budgets for telecommunications infrastructure, particularly for large-scale or multi-site deployments (Smith & Johnson, 2020). While these percentages may appear small relative to overall budgets, their implications are far-reaching when applied across the industry. With millions of users relying on telecommunications services, the costs are inevitably passed down to consumers through higher fees and service charges. For instance, a modest $5 increase in monthly fees per customer to offset NEPA compliance costs translates to $60 annually per consumer—a significant burden for households already grappling with tight budgets and rising living expenses.

The cumulative effect of these added costs is particularly concerning in the context of essential connectivity services. Households in low-income brackets, which often have fewer options for affordable internet and phone services, feel the financial strain more acutely. The additional costs associated with NEPA compliance exacerbate the digital divide by making telecommunications services less accessible to vulnerable populations. This issue is particularly pronounced in rural areas, where infrastructure deployment is already costly and less economically viable for providers. The added financial burden of environmental reviews and tribal clearances further discourages investment in these underserved regions, perpetuating disparities in access to high-speed connectivity (FCC, 2023).

Furthermore, the long-term financial impact of cumulative compliance costs extends beyond consumers. Telecommunications companies must balance the need to recover these expenses with their broader goals of expanding networks and delivering innovative services. The diversion of funds toward regulatory compliance reduces resources available for research, development, and technological advancement, potentially slowing the pace of innovation in critical areas like 5G and smart technologies. This slowdown impacts not only individual users but also industries that rely on advanced telecommunications for growth and competitiveness, including healthcare, education, and commerce.

Addressing the cumulative costs of NEPA compliance requires a multifaceted approach. Standardizing fees for cultural and environmental reviews, streamlining regulatory processes, and implementing categorical exclusions for low-impact projects are critical steps toward reducing financial inefficiencies. Such reforms would alleviate the economic burden on telecommunications providers and consumers alike, ensuring that essential services remain affordable and accessible while maintaining NEPA's commitment to environmental protection and cultural preservation.

How These Costs Are Buried in Carrier Fees and Broader Pricing Structures

Consumers rarely encounter a line item on their phone bills explicitly labeled as "NEPA compliance," yet the associated costs are seamlessly embedded within broader pricing structures and carrier fees. Telecommunications companies strategically distribute these expenses across their customer base to mitigate the financial impact on their operations, ensuring that the burden is less visible but no less significant. For instance, charges such as administrative fees, network improvement surcharges, and other vaguely defined billing categories often conceal NEPA-related compliance costs. These charges are presented as standard operational expenses, obscuring the extent to which regulatory compliance contributes to higher monthly bills (FCC, 2023).

The lack of transparency in how NEPA costs are passed on to consumers compounds the issue. Most customers are unaware that their payments are partially allocated to cover environmental assessments, tribal clearance letters, and project delays tied to NEPA compliance. This opacity prevents consumers from fully understanding the financial implications of regulatory processes on their household budgets. For example, a customer paying an additional $3 to $5 per month due to these hidden costs may not immediately recognize the connection between their bill and the administrative burdens tied to deploying or upgrading telecommunications infrastructure. Over the course of a year, however, these small charges accumulate, adding substantial expenses for households already navigating tight financial margins (Smith & Johnson, 2020).

Additionally, the competitive nature of the telecommunications market discourages companies from explicitly highlighting the costs of regulatory compliance, such as those related to NEPA. Providers fear that openly attributing higher fees to regulatory mandates could alienate customers or tarnish their brand reputation. Instead, these costs are absorbed into complex pricing models that make it difficult for consumers to discern how much of their bill is tied to compliance versus actual service improvements or network investments. This approach shields companies from customer backlash but perpetuates the cycle of inflated costs, as the true financial burden of NEPA-related inefficiencies remains hidden from public scrutiny.

This lack of visibility also undermines accountability and reform efforts. Without clear data on how regulatory expenses affect pricing, consumers and policymakers have little basis for questioning the necessity or efficiency of NEPA compliance processes. As a result, the telecommunications industry continues to absorb and pass on these costs without significant pushback or incentives to streamline regulatory practices. The result is a system where consumers, often unknowingly, fund the inefficiencies and redundancies associated with regulatory overreach, paying more for services that are delayed or incrementally improved due to compliance-related barriers (CEQ, 2020).

To address these issues, greater transparency in billing practices is essential. Clearer communication from telecommunications companies regarding the specific impact of regulatory compliance on pricing could foster consumer awareness and advocacy for reform. Similarly, policymakers could encourage standardized reporting requirements for NEPA-related expenses, enabling both companies and consumers to better understand the financial implications of compliance. These steps could help break the cycle of hidden costs, paving the way for a more balanced approach to regulatory oversight that prioritizes efficiency and affordability alongside environmental and cultural preservation.

Chapter 7

Industry Perspectives

The National Environmental Policy Act (NEPA) plays a pivotal role in regulating federally linked projects, but its application to the telecommunications sector has sparked significant debate. To better understand the impacts of NEPA on this industry, perspectives from telecommunications companies, environmental consultants, and legal experts reveal a landscape fraught with tension between compliance requirements, environmental oversight, and the rapid demands of technological advancement. While NEPA was designed to safeguard the environment from large-scale federal projects, its broad application to low-risk telecommunications infrastructure, such as cell towers and colocations, has created inefficiencies that strain both the industry and consumers. These diverse viewpoints highlight the urgent need for reform to balance environmental protection with the practical realities of modern infrastructure development.

Interviews with Telecommunications Companies, Environmental Consultants, and Legal Experts

To fully grasp the multifaceted impacts of the National Environmental Policy Act (NEPA) on the telecommunications sector, interviews were conducted with key stakeholders, including representatives from telecommunications companies, environmental consultants, and legal experts. These perspectives reveal a complex interplay between regulatory compliance, environmental protection, and the rapid advancement of telecommunications infrastructure.

Telecommunications companies voiced significant frustration over the financial and temporal burdens imposed by NEPA compliance, especially for projects classified as low-risk. One company executive described the process as a "significant bottleneck," explaining that "the process adds months to timelines and thousands of dollars to budgets for projects like small cell installations, which have negligible environmental impact" (FCC, 2023). This delay, the executive argued, slows the deployment of critical infrastructure, such as 5G networks, which rely on dense networks of small cell installations. The prolonged review process not only increases costs but also hampers the industry's ability to meet growing consumer demand for high-speed, reliable connectivity. Additionally, these delays hinder innovation in areas dependent on advanced telecommunications, such as autonomous vehicles and smart cities, creating broader economic and social repercussions.

Environmental consultants offered a contrasting perspective, defending the necessity of NEPA reviews as a means of ensuring due diligence in federally linked projects. One consultant stressed that "even low-risk projects can have cumulative impacts that need to be considered, especially in areas with historical or ecological sensitivity." For example, they pointed out that small cell installations in urban areas may have a minimal environmental footprint individually, but their collective presence could affect local ecosystems or contribute to urban clutter if not properly managed. However, consultants acknowledged that NEPA's application to telecommunications projects often results in inefficiencies. They conceded that clearer guidelines and exemptions for urban or low-impact projects would streamline the process, reduce delays, and focus attention on projects with genuine environmental concerns (Smith & Johnson, 2020).

Legal experts weighed in on the regulatory framework underpinning NEPA, emphasizing the litigation risks that compel companies to undertake exhaustive reviews. An attorney specializing in environmental law explained, "Telecommunications companies often opt for exhaustive reviews to preempt potential lawsuits, even when the likelihood of significant environmental impact is minimal. This practice, while protective, significantly inflates project costs and timelines" (CEQ, 2020). Legal experts highlighted the challenge of balancing the need for compliance with the risk of litigation, particularly in a regulatory landscape where the definition of "significant impact" is often subjective and open to interpretation. They further noted that the lack of clear thresholds for environmental assessments contributes to the overapplication of NEPA, as agencies and developers adopt a "better safe than sorry" approach to avoid costly legal disputes.

These interviews underscore the diverse perspectives on NEPA's role in the telecommunications sector. While telecommunications companies see it as a costly obstacle to infrastructure deployment, environmental consultants and legal experts emphasize its importance in maintaining accountability and preventing environmental oversight from being overlooked. However, all stakeholders agreed that the current application of NEPA could benefit from targeted reforms. Streamlining the process for low-impact projects, clarifying thresholds for significant impact, and implementing categorical exclusions in urban areas were among the suggestions to balance environmental protection with the industry's need for efficiency and growth.

Insights into How the Industry Views NEPA's Role in the Cell Tower Approval Process

The telecommunications industry holds mixed views on the National Environmental Policy Act (NEPA), seeing it as both a vital regulatory tool and a significant obstacle to progress. Telecommunications companies, in particular, contend that NEPA plays a critical role in safeguarding sensitive environments but is often misapplied in low-impact and urban contexts. One industry representative voiced frustration, stating, "NEPA was designed for highways and power plants, not small cell antennas on streetlights" (FCC, 2023). This perspective reflects a broader concern that the law's one-size-fits-all application fails to account for the minimal environmental risks posed by many telecommunications projects. As a result, companies face unnecessary delays and costs that they argue provide little environmental benefit, particularly in urban areas where infrastructure already exists.

Environmental consultants, while defending NEPA's foundational purpose, acknowledged the inefficiencies in its application to the telecommunications sector. One consultant emphasized that NEPA's procedural rigor remains essential for ensuring accountability and public participation in federal projects, stating, "NEPA's intent is valid, but its execution needs modernization" (Smith & Johnson, 2020). Consultants pointed out that NEPA reviews often uncover potential issues that might otherwise be overlooked, particularly in ecologically or culturally sensitive areas. However, they also conceded that applying the same level of scrutiny to minor upgrades or low-risk projects creates unnecessary burdens for both the industry and the agencies tasked with oversight. "The industry would benefit from targeted reforms that streamline the process without compromising environmental integrity," the consultant added, suggesting that improvements in efficiency and clarity could maintain NEPA's effectiveness while reducing its perceived overreach.

Legal experts focused on the balance between compliance and risk management, emphasizing that the real issue lies in how NEPA is implemented rather than the law itself. They noted that NEPA provides robust environmental oversight, but its expansive interpretation by federal agencies often results in overreach, leading to inflated costs and project delays. One legal expert remarked, "Clearer thresholds for triggering reviews could alleviate much of the tension between compliance and progress" (CEQ, 2020). They highlighted that many telecommunications companies conduct exhaustive reviews for low-risk projects to avoid potential litigation, even when environmental impacts are minimal. This approach, while protective, exacerbates the inefficiencies that critics attribute to NEPA.

The perspectives of these stakeholders highlight the complexity of NEPA's role in the telecommunications industry. While its foundational principles of environmental protection and public accountability are widely supported, its application often generates tension between regulatory compliance and infrastructure development. Targeted reforms—such as establishing clearer thresholds for reviews, streamlining processes for low-risk projects, and focusing on substantive rather than procedural compliance—could address these concerns. Such changes would allow NEPA to continue fulfilling its critical environmental mission while reducing unnecessary burdens on the telecommunications sector and, by extension, consumers.

Discussion of Efforts to Streamline the Process Through Exemptions and Reforms

Efforts to streamline NEPA's application in the telecommunications sector have centered on achieving a balance between regulatory compliance and the need for rapid infrastructure development. One widely discussed solution involves implementing categorical exclusions (CEs) for low-impact projects, such as small cell installations on existing structures. These exclusions, already used in other sectors for projects with minimal environmental risk, would allow telecommunications providers to bypass extensive Environmental Assessments (EAs) or Environmental Impact Statements (EISs), significantly cutting costs and reducing delays. For instance, installing antennas on pre-existing utility poles in urban areas could qualify for such exemptions, given the negligible environmental impact of these projects (FCC, 2023). By focusing resources on higher-risk projects, CEs would streamline NEPA compliance without compromising its core purpose of environmental protection.

Another area of focus has been standardizing tribal consultation fees and creating a centralized system to manage these interactions. Current practices often involve inconsistent fee structures and protracted negotiations, which inflate costs and delay projects unnecessarily. Industry advocates propose that the Federal Communications Commission (FCC) or another neutral entity oversee these interactions, establishing clear guidelines for fair compensation and eliminating redundancies. This approach would improve transparency and reduce administrative burdens for both developers and tribes, fostering a more collaborative process. "Streamlining the process doesn't mean eliminating oversight," a legal expert clarified. "It's about focusing resources where they're needed most—on projects with genuine environmental risks" (Smith & Johnson, 2020).

In addition to these procedural reforms, telecommunications companies have engaged directly with policymakers to advocate for broader regulatory modernization. Collaborative efforts with the Council on Environmental Quality (CEQ) aim to update NEPA regulations to reflect the realities of modern telecommunications infrastructure. Companies are also partnering with tribal leaders to develop fair and consistent consultation practices that honor cultural preservation while avoiding excessive fees and delays. Legal experts have suggested that the FCC take a more active role in mediating disputes and clarifying review requirements, reducing the likelihood of unnecessary or redundant evaluations. This would ensure that NEPA compliance remains effective and efficient, aligning with its original intent.

Industry representatives express cautious optimism about these proposed reforms, emphasizing the importance of maintaining NEPA's environmental safeguards while addressing inefficiencies. "We're not asking to bypass environmental laws," one executive explained. "We just need a process that reflects the realities of modern telecommunications infrastructure" (CEQ, 2020). These efforts to modernize NEPA aim to protect sensitive environments and cultural sites while enabling the rapid deployment of critical technologies like 5G, ensuring that regulatory compliance supports rather than hinders technological progress.

Chapter 8

A Comparison with Similar Infrastructure

The regulatory disparity between cell towers and similar infrastructure, such as streetlights, power line poles, and sports field lights, highlights a critical inefficiency in how NEPA is applied. While streetlights and utility poles are installed without extensive environmental reviews, cell towers—even small installations on pre-existing structures—are subject to rigorous assessments due to their connection to federally licensed frequencies. This chapter explores the similarities and differences between these types of infrastructure, shedding light on the disproportionate regulatory burden placed on cell towers. By examining why certain structures are exempt from NEPA compliance and evaluating the arguments for categorizing cell towers similarly, this chapter underscores the need for reforms that align regulatory oversight with actual environmental risks, fostering both efficiency and fairness.

Streetlights, Power Line Poles, and Sports Field Lights: Why These Structures Do Not Require NEPA Reviews

Streetlights, power line poles, and sports field lights are ubiquitous forms of infrastructure that serve essential functions in urban, suburban, and rural environments. These structures are generally exempt from NEPA reviews because their installation and operation lack the federal actions that trigger NEPA compliance. NEPA applies primarily to projects with federal involvement, such as those requiring permits, funding, or licenses from federal agencies. In contrast, the installation of streetlights, power line poles, and sports field lights is typically regulated by state or local authorities, thereby bypassing the federal nexus necessary for NEPA applicability (Council on Environmental Quality [CEQ], 2020).

Additionally, these infrastructure types are considered routine, low-impact, and predictable in their design and installation. Streetlights, for instance, are commonly added to pre-developed urban or suburban areas, where the environmental and cultural risks are minimal. The standardized nature of their designs and the lack of significant ground disturbance reduce the likelihood of adverse impacts, making comprehensive environmental reviews unnecessary. Similarly, power line poles are installed as part of existing utility grids, often replacing or upgrading pre-existing structures, further minimizing environmental disruption.

Sports field lights also exemplify infrastructure exempt from NEPA reviews due to their placement in areas that have already been extensively modified, such as athletic complexes or school facilities. These installations are generally managed by local municipalities or private organizations and do not require federal permits or licensing. This localized regulatory framework allows these projects to avoid the time-consuming and costly processes associated with NEPA compliance, streamlining their approval and implementation (Federal Communications Commission [FCC], 2023). The absence of federal oversight for such low-impact infrastructure underscores the practicality of excluding routine structures from NEPA's scope, allowing resources and attention to focus on projects with potentially significant environmental consequences.

Similarities and Differences Between Cell Towers and These Infrastructure Types

Cell towers share numerous similarities with streetlights, power line poles, and sports field lights in terms of their physical attributes and environmental footprint. All these structures are designed for specific utility functions—streetlights for illumination, power line poles for energy distribution, and cell towers for telecommunications. These installations are typically situated in developed areas, such as urban neighborhoods, suburban streets, or pre-existing utility corridors, where their impact on natural environments is minimal. In urban settings, aesthetic considerations drive the design of many cell towers, making them visually comparable to streetlights or power line poles. Small cell installations for 5G networks have further reinforced this similarity by integrating telecommunications equipment directly onto existing infrastructure like utility poles and streetlights, effectively minimizing additional environmental impacts (Smith & Johnson, 2020).

The key difference between these infrastructure types lies in the federal licensing requirements associated with cell towers. Telecommunications providers require licenses from the Federal Communications Commission (FCC) to operate the radio frequencies used for their networks. This federal involvement triggers the National Environmental Policy Act (NEPA) compliance process, subjecting cell towers to an additional layer of regulatory scrutiny. In contrast, streetlights, power line poles, and sports field lights, which are typically governed by local or state regulations, do not involve federal oversight and are therefore exempt from NEPA reviews.

This distinction imposes significant procedural and financial burdens on cell tower projects that similar infrastructure types avoid. For instance, installing a streetlight or power line pole can be completed in a matter of weeks without the need for environmental reviews. However, a small cell installation on an identical utility pole may face months of delays due to NEPA-mandated assessments and tribal consultations, even in cases where environmental risks are negligible (Federal Communications Commission [FCC], 2023). These inconsistencies highlight the disproportionate regulatory treatment of cell towers compared to other utility structures, raising questions about the practicality of subjecting telecommunications projects to NEPA's broad scope in low-impact scenarios.

Arguments for Categorizing Cell Towers Similarly to These Exempt Structures

The argument for exempting cell towers and the colocation of cell antennas from NEPA compliance is rooted in the minimal environmental impact of these projects, particularly in urban and suburban settings. These installations often involve integrating telecommunications equipment onto existing infrastructure, such as utility poles or buildings, or constructing small-scale towers with limited physical and ecological footprints. Proponents of regulatory reform argue that such projects share the low-risk profile of streetlights, power line poles, and sports field lights—none of which are subject to NEPA reviews (CEQ, 2020). This inconsistency in regulatory treatment highlights the inefficiency of applying NEPA to cell towers, especially when it adds significant costs and delays without yielding meaningful environmental benefits.

Categorizing small cell installations and antenna colocations as routine infrastructure would streamline the regulatory process by removing the requirement for Environmental Assessments (EAs) or Environmental Impact Statements (EISs) for projects with negligible risks. Extending categorical exclusions (CEs) to cover these installations in developed areas is a logical step toward aligning federal regulations with practical realities. By focusing NEPA compliance on projects with genuine potential for significant environmental or cultural impacts, federal agencies could better allocate resources and expedite the deployment of critical telecommunications infrastructure (Smith & Johnson, 2020).

Critics of the current application of NEPA to cell towers emphasize that the law was originally intended to address large-scale projects with substantial environmental risks, such as highways, dams, or mining operations. Subjecting low-impact telecommunications projects to the same level of scrutiny dilutes NEPA's purpose, creating a bureaucratic burden that slows progress in areas like 5G network deployment. This inefficiency is particularly problematic given the urgent need for improved connectivity to support modern technologies, such as smart cities and the Internet of Things (IoT). The financial and temporal costs of NEPA compliance are ultimately borne by consumers, who face higher service fees and slower access to next-generation connectivity (FCC, 2023).

Reclassifying cell towers and small cell installations under a similar regulatory framework as streetlights and power line poles would reduce these inefficiencies while maintaining NEPA's core principles. Such a change would reflect the realities of modern telecommunications infrastructure, where the majority of installations occur in areas with pre-existing development and minimal environmental sensitivity. Critics of reform often argue that exemptions could weaken environmental oversight, but proponents counter that reforms would not eliminate NEPA compliance altogether. Instead, they would focus environmental reviews on high-risk projects, ensuring that NEPA remains an effective tool for addressing substantial environmental concerns (Smith & Johnson, 2020).

By adopting a targeted approach to NEPA compliance for telecommunications infrastructure, federal agencies could achieve a balance between environmental protection and technological progress. Streamlined regulations would reduce project delays, lower costs, and support the rapid deployment of advanced telecommunications networks, all while preserving NEPA's original intent. Such reforms would ensure that the law continues to serve its critical purpose without becoming an obstacle to innovation and economic growth.

The extensive application of NEPA to telecommunications infrastructure, particularly in urban areas, on existing buildings, and for minor tower extensions, represents a clear example of regulatory overreach that comes at the expense of consumers. Having personally worked on over 10,000 cell towers and collocations where NEPA compliance was mandated, I have seen firsthand the inefficiencies and unnecessary costs imposed by applying these regulations to low-risk projects. While NEPA serves an essential purpose in protecting the environment, its blanket application to projects with negligible impact dilutes its effectiveness and inflates costs for both industry and consumers. Each month, cellular users unknowingly pay for these inefficiencies through their phone bills, funding redundant reviews that yield little to no environmental benefit. It is imperative that regulatory frameworks evolve to reflect the realities of modern telecommunications, focusing resources on projects with genuine environmental or cultural risks and alleviating unnecessary burdens on consumers and the industry alike. Reforming NEPA's application to cell towers is not just about reducing costs—it's about restoring balance and purpose to one of the nation's most important environmental laws.

Chapter 9

Reforming NEPA for the Modern Era

The National Environmental Policy Act (NEPA), a cornerstone of environmental protection, was designed to address the significant ecological risks posed by large-scale federal projects. However, as the landscape of infrastructure development evolves, NEPA's application has struggled to keep pace with modern technologies and priorities. Nowhere is this more evident than in the telecommunications sector, where low-impact projects, such as small cell installations and tower modifications, are subjected to the same regulatory burdens as major construction initiatives. This disconnect has led to inefficiencies, inflated costs, and delays that hinder technological progress and impose unnecessary financial burdens on consumers. Reforming NEPA for the modern era requires a thoughtful recalibration of its processes, focusing on streamlining compliance for low-risk projects while maintaining robust protections for genuinely significant environmental and cultural concerns. By adapting NEPA to the realities of today's infrastructure needs, we can preserve its original intent while fostering innovation and connectivity.

Proposals to Streamline NEPA for Telecommunications Projects

Streamlining the National Environmental Policy Act (NEPA) for telecommunications projects has become a pressing need in an era of rapid technological advancement and increasing demand for robust wireless connectivity. While NEPA serves as a critical tool for safeguarding the environment, its blanket application to both high-impact and low-impact projects has created inefficiencies that hinder infrastructure deployment. Small cell installations, routine upgrades, and colocations—key components of modern telecommunications networks—often face the same regulatory scrutiny as large-scale infrastructure projects, despite their minimal environmental impact. Proposals to reform NEPA for telecommunications aim to strike a balance between preserving environmental protections and accelerating the rollout of critical infrastructure, ensuring that compliance processes are proportionate, efficient, and reflective of contemporary needs.

Categorical Exclusions for Low-Impact Projects

One of the most effective proposals for reforming NEPA is the broader application of categorical exclusions (CEs) for low-impact telecommunications projects. CEs are a critical tool within NEPA that allow projects with minimal environmental risks to bypass extensive Environmental Assessments (EAs) or Environmental Impact Statements (EISs), significantly streamlining the compliance process (Council on Environmental Quality [CEQ], 2020). For telecommunications infrastructure, this reform could be transformative, particularly for small cell installations, antenna upgrades, or colocations on existing structures in urban or suburban areas. These projects, which typically have negligible environmental impact, are currently subjected to the same rigorous reviews as large-scale developments such as new tower constructions in pristine natural habitats.

The inefficiency of applying NEPA to these low-impact projects is evident in both time and cost. For example, small cell installations, critical for the rollout of 5G networks, are often located on pre-existing infrastructure like utility poles or streetlights in developed areas. Despite their minimal environmental footprint, these projects face procedural hurdles that add months to approval timelines and significantly increase project budgets. The redundancy of this process not only delays the deployment of essential infrastructure but also inflates costs that are ultimately passed on to consumers (FCC, 2023).

Expanding CEs to include routine telecommunications installations offers a practical solution. Federal agencies could establish clear and consistent thresholds for CE eligibility, focusing on projects that reuse existing infrastructure or are located in already-developed urban or suburban areas. For instance, colocations—where new antennas are added to pre-existing towers—are low-risk projects that could benefit greatly from this streamlined approach. By removing unnecessary procedural requirements, agencies could prioritize resources for high-impact projects where environmental reviews are truly needed.

Moreover, clear guidelines for CE eligibility could help prevent regulatory bottlenecks and ensure consistency across jurisdictions. For example, exemptions could specify that projects in urban areas, where the potential for environmental or cultural impacts is negligible, automatically qualify for CEs. This would not only expedite the approval process but also reduce administrative burdens for federal agencies, developers, and stakeholders alike.

Importantly, expanding CEs for telecommunications projects does not compromise NEPA's foundational goal of environmental protection. By focusing reviews on projects with significant risks—such as those in sensitive ecological areas or near cultural heritage sites—agencies can ensure that resources are allocated where they are most needed. This targeted approach would enhance the effectiveness of NEPA, preserving its intent while adapting it to the realities of modern infrastructure development.

Incorporating expanded CEs into NEPA reform is a balanced and pragmatic step forward. It aligns with the original spirit of the law by protecting the environment while addressing inefficiencies that hinder critical infrastructure deployment. As policymakers consider the future of NEPA, emphasizing the use of CEs for low-impact telecommunications projects offers a pathway to more efficient and effective regulatory oversight, fostering both innovation and environmental stewardship.

Differentiating Between Significant and Negligible Environmental Impacts

A pivotal proposal for NEPA reform involves redefining the concept of "significant impact" to reduce the number of unnecessary reviews for low-risk projects. Under current NEPA guidelines, the term "significant impact" remains vague, allowing for overly cautious interpretations that lead to exhaustive Environmental Assessments (EAs) and Environmental Impact Statements (EISs), even for projects with negligible environmental risks (Smith & Johnson, 2020). This lack of clarity incentivizes consulting firms and federal agencies to err on the side of caution, often opting for full reviews to preempt potential litigation or criticism. While this approach may mitigate legal risks, it contributes to inefficiencies that inflate costs and delay infrastructure development.

To streamline the NEPA process, it is essential to establish clearer distinctions between significant and negligible impacts. By providing more explicit criteria, federal agencies can better allocate resources, ensuring that detailed reviews are reserved for projects with genuine environmental or cultural risks. For instance, small cell installations, which are typically deployed on existing infrastructure like streetlights or utility poles in urban or suburban areas, could be explicitly classified as negligible-impact projects. This classification would exempt them from the extensive procedural requirements currently imposed under NEPA, significantly reducing costs and deployment timelines.

The benefits of redefining "significant impact" extend beyond cost savings and expedited timelines. This change would align NEPA's application with its original purpose—evaluating projects with substantial environmental consequences. By focusing resources on high-risk projects, agencies can maintain rigorous environmental oversight while reducing procedural burdens on low-impact initiatives. For example, projects in environmentally sensitive areas, such as wetlands or habitats for endangered species, would still undergo comprehensive reviews, preserving NEPA's foundational goals of environmental protection and accountability (Council on Environmental Quality [CEQ], 2020).

Critics of NEPA's current application argue that the law's broad interpretation has diluted its effectiveness, diverting attention and resources from critical environmental issues to procedural compliance for low-impact projects. For telecommunications companies, this misalignment disproportionately affects the deployment of essential infrastructure, such as 5G networks, where small cell installations are integral to expanding coverage and connectivity. By redefining "significant impact," NEPA could better balance environmental protection with the need for rapid technological advancement (Federal Communications Commission [FCC], 2023).

Policymakers could implement this reform by developing a tiered system for NEPA reviews, categorizing projects based on their potential environmental impact. For negligible-impact projects, such as small cell installations in urbanized areas, categorical exclusions could apply automatically, bypassing the need for EAs or EISs. Medium-impact projects might require a simplified EA process, while high-impact projects would continue to undergo full EIS reviews. This stratified approach would enhance NEPA's efficiency and effectiveness, reducing costs and delays for low-risk projects while preserving the integrity of reviews for significant developments.

Redefining "significant impact" is a necessary evolution of NEPA that reflects the realities of modern infrastructure needs. By providing clear and actionable guidelines, this reform would reduce regulatory bottlenecks, empower federal agencies to prioritize meaningful environmental oversight, and support the rapid deployment of critical telecommunications infrastructure. Ultimately, this targeted approach would benefit both industry stakeholders and consumers, fostering innovation while upholding NEPA's core principles.

Examples of Successful Regulatory Reforms in Other Industries

Across various industries, regulatory reforms have demonstrated that balancing oversight with efficiency is achievable without compromising critical safeguards. By addressing inefficiencies and adopting targeted changes, sectors such as transportation, energy, and construction have successfully streamlined compliance processes while maintaining their commitment to safety and environmental protection. These examples offer valuable lessons for telecommunications, where NEPA compliance has created significant delays and financial burdens. Industries that once faced similar regulatory hurdles have implemented reforms such as categorical exclusions, standardized review procedures, and technological innovations to reduce costs and accelerate project timelines. Examining these successes provides a roadmap for modernizing NEPA's application in telecommunications while preserving its core objectives of accountability and environmental stewardship.

The Energy Sector: Streamlining for Renewable Projects

The energy industry offers a compelling example of how regulatory reforms can balance environmental protection with the need for expedited project deployment. In 2020, the Trump administration implemented changes to NEPA that focused on reducing inefficiencies in the approval process for renewable energy projects. These reforms included strict deadlines for completing Environmental Assessments (EAs) and Environmental Impact Statements (EISs), often criticized for their protracted timelines, as well as the expanded use of categorical exclusions (CEs) for projects with minimal environmental risks. Solar, wind, and battery storage installations on federal lands particularly benefited from these streamlined processes, enabling faster deployment of renewable energy infrastructure while maintaining environmental safeguards (U.S. Department of the Interior, 2020). These changes reflect a shift toward a more practical application of NEPA that prioritizes the timely development of projects deemed beneficial for societal and environmental goals.

The telecommunications sector could benefit from adopting a similar approach. Just as renewable energy projects are crucial for reducing carbon emissions and advancing sustainable practices, telecommunications infrastructure plays a vital role in enhancing public connectivity, supporting economic growth, and facilitating technological innovation. Prolonged delays in deploying critical infrastructure like 5G networks diminish these potential benefits, leaving communities with slower connectivity and outdated services. By introducing reforms similar to those in the energy sector—such as CEs for small cell installations or co-located equipment upgrades—telecommunications projects could avoid unnecessary procedural bottlenecks while still adhering to NEPA's overarching goals of accountability and environmental protection.

Furthermore, the energy industry's reforms underscore the importance of aligning regulatory frameworks with the evolving needs of modern infrastructure. Just as renewable energy technologies required a reassessment of NEPA's application to accommodate their unique characteristics, telecommunications projects need tailored solutions to reflect their typically low environmental impact. This alignment could pave the way for more efficient infrastructure development, ensuring that federal oversight focuses on high-impact projects while enabling swift progress in areas that offer substantial public benefits with minimal risks.

Transportation Sector Reforms

The transportation industry provides valuable insights into how NEPA reforms can streamline regulatory processes without compromising environmental protections. The Fixing America's Surface Transportation (FAST) Act of 2015 introduced provisions specifically aimed at expediting environmental reviews for transportation infrastructure projects. A central element of these reforms was the establishment of the "FAST-41" framework, which identifies high-priority infrastructure projects and ensures their reviews are conducted within defined timeframes. This process requires federal agencies to coordinate efforts, set timelines, and establish accountability measures, reducing delays and providing developers with greater certainty in project planning and execution (Council on Environmental Quality [CEQ], 2020).

The FAST Act reforms have been particularly effective in addressing inefficiencies for projects that are crucial for public infrastructure. By prioritizing and accelerating reviews, the framework has allowed developers to complete critical transportation initiatives—such as bridges, highways, and rail projects—more efficiently. For example, the act's emphasis on interagency coordination and transparency has streamlined approvals for projects with clear public benefits, while maintaining robust environmental oversight where needed.

The telecommunications sector could adopt similar prioritization systems to expedite NEPA compliance for critical projects, particularly those aimed at closing the digital divide in underserved communities. High-priority telecommunications projects, such as 5G network deployment in rural or economically disadvantaged areas, could benefit from the creation of a "telecommunications FAST" framework. This approach could include stricter timelines for Environmental Assessments (EAs) and Environmental Impact Statements (EISs), as well as mechanisms for resolving interagency bottlenecks. By targeting NEPA reforms toward projects that provide significant societal benefits with minimal environmental risks, such a framework could accelerate infrastructure deployment without sacrificing environmental protections.

Furthermore, the transportation industry's experience highlights the importance of tailoring NEPA processes to reflect the realities of modern infrastructure. Just as the FAST Act addressed the unique challenges of large-scale transportation projects, telecommunications regulations must evolve to accommodate the rapid deployment of small cell installations and other low-impact infrastructure critical to 5G networks. Categorical exclusions (CEs) for routine projects, clearer definitions of "significant impact," and standardized consultation procedures could help ensure that NEPA compliance focuses on areas of genuine concern rather than creating unnecessary delays and costs.

Reforming NEPA to reflect the modern telecommunications landscape is not just a matter of efficiency—it is essential for balancing environmental oversight with the urgent need for expanded connectivity. Drawing lessons from successful reforms in the transportation industry, the telecommunications sector can adopt best practices that reduce costs and delays while preserving NEPA's mission of environmental stewardship. These reforms would ensure that NEPA remains a relevant and effective tool in a rapidly changing regulatory environment, enabling critical infrastructure projects to move forward without undue obstacles.

Douglas B Sims, PhD

Chapter 10

Balancing Environmental Protection and Consumer Needs

As the telecommunications industry evolves to meet the growing demand for high-speed connectivity, particularly with the deployment of 5G networks, balancing environmental protection with consumer affordability has become a critical challenge. The National Environmental Policy Act (NEPA) plays a pivotal role in safeguarding ecosystems and cultural resources during infrastructure development. However, its application to low-impact projects often leads to unnecessary bureaucracy, inflating costs and delaying progress. These expenses, passed on to consumers through higher service fees, disproportionately affect households already facing financial pressures. This chapter explores strategies for addressing legitimate environmental concerns while streamlining regulatory processes, leveraging technological advancements, and ensuring that consumer needs for affordable, reliable telecommunications services remain at the forefront. By reimagining how NEPA is applied, it is possible to strike a balance that protects the environment without imposing undue burdens on consumers.

Ensuring That Legitimate Environmental Concerns Are Addressed Without Unnecessary Bureaucracy.

Balancing environmental protection with infrastructure development requires a regulatory framework that prioritizes meaningful oversight while reducing inefficiencies and unnecessary delays. Legitimate environmental concerns, such as preserving biodiversity, safeguarding culturally significant sites, and preventing ecosystem degradation, are the cornerstones of the NEPA process. These priorities reflect society's commitment to ensuring that development does not come at the cost of irreparable harm to natural and cultural resources. However, the current application of NEPA often imposes bureaucratic hurdles even for low-impact projects, diverting resources from genuine environmental oversight to redundant and costly procedures (Council on Environmental Quality [CEQ], 2020).

The telecommunications sector exemplifies the challenges posed by this regulatory overreach. For example, small-scale projects like 5G small cell installations, which are typically mounted on existing structures such as utility poles or streetlights in urban areas, frequently undergo the same rigorous NEPA reviews as large-scale rural developments. Despite their negligible environmental footprint, these projects are subject to extensive Environmental Assessments (EAs) or Environmental Impact Statements (EISs), consuming significant time and financial resources without corresponding environmental benefits. This one-size-fits-all approach not only delays critical infrastructure deployment but also undermines NEPA's original purpose of addressing significant environmental risks.

To address these inefficiencies, the NEPA process could benefit from reforms that categorize projects based on their potential environmental impact. Establishing clear thresholds for when EAs and EISs are required would ensure that regulatory resources are directed toward projects with genuine risks, such as those in ecologically sensitive areas or near culturally significant sites. Projects with minimal potential for adverse impacts, such as colocations on existing cell towers or small cell installations in urbanized areas, could be designated as low-risk and exempt from extensive reviews through the implementation of categorical exclusions (CEs). CEs are already widely used in other industries, such as transportation and energy, to streamline approvals for routine, low-impact projects. Applying a similar approach in the telecommunications sector would reduce unnecessary delays and costs while maintaining environmental protections.

Policymakers could also introduce more targeted reforms to NEPA that align its application with the realities of contemporary infrastructure needs. For instance, updating federal guidelines to explicitly define what constitutes a "significant impact" would reduce ambiguity and prevent unnecessary reviews for projects with negligible risks. By establishing clearer criteria, agencies could focus their efforts on projects that genuinely warrant comprehensive environmental evaluations, ensuring that NEPA remains a tool for meaningful oversight rather than a procedural bottleneck (Federal Communications Commission [FCC], 2023).

These reforms would not only streamline the regulatory process but also address the broader societal implications of NEPA's inefficiencies. Delays in telecommunications projects, particularly those related to 5G deployment, hinder technological advancement and leave consumers paying higher costs for suboptimal connectivity. By modernizing NEPA to reflect the realities of low-impact projects, policymakers can create a regulatory system that balances environmental protection with the urgent need for affordable and reliable telecommunications infrastructure. This approach ensures that resources are allocated where they are most needed, preserving NEPA's core mission while fostering innovation and connectivity.

Strategies to Protect the Environment While Reducing Costs for Consumers.

Protecting the environment while minimizing costs is achievable through strategic reforms and innovative approaches. These goals can complement each other when efficiency and targeted oversight are prioritized. A key strategy involves leveraging existing infrastructure to avoid unnecessary land disturbance and reduce the need for new construction. Colocating new antennas on existing towers, utility poles, or streetlights minimizes physical and environmental impacts, bypassing the need for extensive land clearing or excavation. This practice not only reduces environmental risks but also eliminates the requirement for exhaustive Environmental Assessments (EAs) or Environmental Impact Statements (EISs) for projects with negligible impacts (Smith & Johnson, 2020). Furthermore, such initiatives align with modern urban planning principles that emphasize optimizing existing resources to meet growing demands.

Another critical strategy is addressing the inefficiencies associated with tribal consultations, particularly the inconsistent and excessive fees for clearance letters. Telecommunications projects often encounter inflated costs due to varying tribal fees, which can range from a few thousand dollars to tens of thousands for similar projects. These inconsistencies create financial unpredictability and unnecessarily inflate project budgets. Implementing standardized fees would ensure fair compensation for tribes while preventing financial exploitation. A transparent fee structure could also improve relations between developers and tribal organizations by fostering trust and mutual understanding, streamlining the overall consultation process (Federal Communications Commission [FCC], 2023).

Policymakers could further enhance efficiency by introducing digital platforms to centralize environmental and cultural review processes. Currently, developers must navigate a fragmented system involving multiple agencies and stakeholders, leading to administrative inefficiencies and delays. A centralized platform would provide a unified interface for managing NEPA reviews, tribal consultations, and other compliance requirements. Developers could submit necessary documents, track project progress, and engage with reviewing bodies in real-time, reducing bureaucratic bottlenecks and miscommunications. Such a system would also enhance transparency, allowing all stakeholders—including federal agencies, tribes, and developers—to operate within a clear, consistent framework. By expediting decision-making and reducing administrative burdens, these digital innovations could significantly lower compliance costs, ultimately benefiting consumers through reduced service fees.

These strategies demonstrate that environmental protection and cost minimization are not opposing goals. By emphasizing the use of existing infrastructure, reforming tribal consultation processes, and leveraging technology to streamline compliance, policymakers and industry stakeholders can create a balanced regulatory framework. Such reforms would uphold NEPA's core principles while addressing its inefficiencies, ensuring both environmental integrity and the affordability of telecommunications services.

The Role of Technology in Minimizing Cell Tower Impacts.

Advances in technology are transforming the telecommunications industry, offering innovative solutions to reduce the environmental impact of infrastructure development. One of the most significant advancements is the proliferation of smaller cell sites, particularly for 5G networks. Unlike traditional macro towers, these small cells have a much smaller footprint and can be seamlessly integrated into existing infrastructure, such as streetlights, utility poles, and building facades. By using existing structures, small cells minimize both environmental and aesthetic impacts, reducing the need for land clearing and extensive construction activities (Smith & Johnson, 2020). These installations are particularly well-suited for urban areas, where infrastructure is already dense, further mitigating potential disruptions to natural ecosystems.

Another promising avenue is the adoption of shared infrastructure. By collaborating to use shared towers and sites, telecommunications providers can significantly reduce the need for redundant installations in the same geographic area. Shared infrastructure, such as neutral-host towers that accommodate multiple carriers, not only decreases land use but also streamlines NEPA compliance. Consolidating environmental reviews for shared sites reduces duplication and administrative burdens while maintaining robust oversight. For instance, neutral-host providers manage shared infrastructure, enabling multiple carriers to utilize a single site, which promotes efficiency and sustainability. This approach also lowers overall deployment costs, creating financial benefits that can be passed on to consumers (Federal Communications Commission [FCC], 2023).

Emerging technologies such as smart sensors, predictive analytics, and geographic information systems (GIS) further enhance the ability to minimize environmental risks. These tools provide precise data for site selection, helping companies identify areas of low environmental and cultural sensitivity. For example, GIS and remote sensing technologies can map ecosystems, wetlands, and other sensitive areas, ensuring that new infrastructure avoids high-impact zones. Predictive analytics can also model potential environmental impacts, enabling developers to preemptively address issues and optimize project planning. By leveraging these technologies, telecommunications companies can ensure that infrastructure is deployed responsibly, balancing connectivity needs with environmental preservation.

Integrating these technological innovations into the regulatory framework is essential for modernizing NEPA compliance. Federal agencies, in collaboration with the telecommunications industry, can adopt these tools to streamline reviews and reduce the environmental footprint of new installations. For example, agencies could encourage or require the use of GIS-based site assessments as part of the NEPA process, ensuring that projects prioritize low-impact locations. Additionally, shared infrastructure initiatives could be incentivized through grants or regulatory exemptions, further promoting collaboration among providers.

By embracing smaller cell sites, shared infrastructure, and advanced technologies, the telecommunications industry can achieve a balanced approach to infrastructure deployment. These innovations not only reduce environmental impacts but also enhance efficiency and sustainability. Such a strategy ensures that the benefits of expanded connectivity and technological progress are realized without compromising the integrity of natural and cultural resources. This alignment of environmental protection with infrastructure growth underscores the potential for technology to transform regulatory processes and deliver solutions that serve both society and the planet.

Douglas B Sims, PhD

Chapter 11

Overreach and the Consumer Burden

The National Environmental Policy Act (NEPA), while vital for safeguarding the environment, has become a double-edged sword in the telecommunications sector. Originally intended to ensure that large-scale federal projects considered their environmental impacts, NEPA's application has expanded to include low-risk projects like small cell installations and minor upgrades to existing towers. This overreach imposes unnecessary regulatory burdens, inflating project budgets and delaying deployment timelines. Telecommunications companies, faced with mounting compliance costs, pass these expenses onto consumers through higher phone bills and surcharges. Meanwhile, environmental consulting firms and other industry players profit from the inefficiencies in the system, perpetuating a cycle of overregulation with limited environmental benefit. This chapter explores how NEPA's misuse disproportionately impacts consumers and highlights the urgent need for reform to balance environmental protections with affordability and efficiency.

How NEPA's Misuse Inflates Phone Bills

The National Environmental Policy Act (NEPA), originally envisioned to address significant environmental concerns associated with large-scale federal projects, has become a tool for overregulation in the telecommunications industry. While its core purpose is to ensure that infrastructure development respects environmental and cultural resources, its application to low-risk projects such as cell tower upgrades or small cell installations has increasingly diverged from its original intent. Projects that pose minimal environmental risks are often subjected to extensive Environmental Assessments (EAs) and Environmental Impact Statements (EISs), processes designed for more impactful undertakings like highways or energy plants. These requirements inflate project budgets, delay timelines, and hinder technological advancements, especially in the deployment of 5G networks, which rely on rapid and dense infrastructure deployment to meet connectivity demands.

The financial consequences of these inefficiencies are significant and widespread. Telecommunications companies, burdened by the excessive costs of compliance, pass these expenses onto consumers, embedding them in service fees and monthly bills. For example, an Environmental Assessment for a small cell installation on a pre-existing utility pole in a developed urban area may cost tens of thousands of dollars and delay the project by months, despite its negligible environmental impact (Smith & Johnson, 2020). The cumulative effect across the industry results in higher phone bills for millions of consumers, disproportionately affecting low-income households that are least equipped to absorb rising costs.

Moreover, NEPA's current implementation creates a misallocation of resources, diverting attention from projects that genuinely threaten sensitive environments to those with minimal or no environmental significance. This misuse not only undermines the spirit of NEPA but also erodes consumer trust, as hidden compliance costs become a burden over which they have no control or transparency. By addressing these inefficiencies and recalibrating NEPA's application to focus on meaningful environmental protections, the telecommunications sector could achieve a balance between regulatory compliance and consumer affordability (Federal Communications Commission [FCC], 2023).

Examples of NEPA Overreach

The implementation of NEPA, while intended to safeguard environmental and cultural resources, has increasingly been applied in ways that exceed its original purpose. This overreach is particularly evident in the telecommunications sector, where projects with minimal environmental impact often face the same rigorous reviews as large-scale infrastructure developments. Examples abound of low-risk initiatives, such as cell tower installations in urban areas or minor modifications to existing infrastructure, being subjected to unnecessary Environmental Assessments (EAs) or Environmental Impact Statements (EISs). These redundant reviews inflate costs, delay critical projects, and offer little environmental benefit. Instead of enhancing protections, this regulatory overreach imposes significant financial and operational burdens on telecommunications companies, which are ultimately passed on to consumers in the form of higher phone bills. Examining specific instances of NEPA overreach highlights the urgent need for reform to restore the law's focus on genuinely impactful projects.

Environmental Reviews for Towers in Developed, Low-Impact Areas

One glaring example of NEPA overreach is the application of full environmental reviews to cell towers and small cell installations in urban or suburban areas where pre-existing infrastructure dominates the landscape. These locations often feature extensive development, such as paved roads, existing utility poles, and dense commercial or residential structures, leaving minimal scope for environmental disruption. Despite this, small antennas added to utility poles or colocated on existing towers are frequently subjected to the same level of scrutiny as new towers constructed in pristine or environmentally sensitive areas. This includes comprehensive Environmental Assessments (EAs) or Environmental Impact Statements (EISs), which involve detailed reviews of potential impacts on wildlife, ecosystems, cultural resources, and local communities.

Such regulatory demands add months of delays to project timelines and impose thousands of dollars in compliance costs, often including consulting fees, public comment processes, and tribal consultations. These requirements fail to yield meaningful environmental benefits in urban or suburban contexts, where the environmental risks associated with small cell installations are negligible due to the pre-developed nature of the sites. For example, an antenna added to a utility pole on a busy city street is unlikely to affect wildlife habitats or cultural landmarks, yet it may still require redundant reviews, delaying deployment of critical infrastructure like 5G networks and inflating project budgets unnecessarily (Council on Environmental Quality [CEQ], 2020).

This approach reflects a misalignment between NEPA's original intent and its current implementation. Originally designed to address large-scale federal projects with potentially significant environmental consequences, NEPA's blanket application to low-impact telecommunications projects undermines its effectiveness. Instead of focusing on genuine environmental risks, resources are wasted on redundant reviews that hinder infrastructure development and increase costs for both the industry and consumers. Reforming NEPA to exclude such projects from excessive scrutiny would not only streamline telecommunications deployment but also restore the law's purpose of protecting truly vulnerable environments.

Duplication of Reviews for Minor Modifications

NEPA requirements frequently mandate new reviews for minor upgrades to existing infrastructure, such as antenna replacements or equipment updates, even when these changes have no substantial effect on the tower's footprint or environmental impact. These upgrades, which are typically routine maintenance or technological improvements, often face the same regulatory hurdles as entirely new installations. Despite the lack of significant environmental risks, consulting firms often recommend redundant reviews to avoid potential litigation or to ensure full compliance with the law. This practice reflects a cautious approach to NEPA's broad and sometimes ambiguous requirements, which can penalize even low-risk projects with time-consuming and costly procedures.

For example, a telecommunications tower that has previously undergone a comprehensive environmental assessment may still require additional reviews for minor modifications. These reviews often duplicate prior evaluations, offering little new information or insight into environmental impacts. This redundancy not only creates inefficiencies but also inflates project budgets unnecessarily, as companies are forced to pay for repeated consulting services and administrative processes (Smith & Johnson, 2020).

The financial burden of these redundant reviews is ultimately passed on to consumers through higher service fees and phone bills. Meanwhile, these delays hinder the rapid deployment of critical telecommunications infrastructure, such as 5G networks, which rely on frequent upgrades to ensure optimal performance and connectivity. Reforming NEPA to exempt minor modifications or streamline the review process for such low-risk projects would reduce unnecessary costs and allow resources to be focused on projects with genuinely significant environmental impacts.

Delays in 5G Deployment Due to Regulatory Excess

The rollout of 5G networks, heralded as a transformative technology for connectivity and innovation, has been significantly impeded by excessive NEPA compliance requirements. Small cell installations, which are critical for 5G functionality due to their ability to handle high data loads with minimal latency, often face the same extensive environmental review processes as new macro towers. These reviews are mandated even when the installations are placed on existing infrastructure such as utility poles, streetlights, or rooftops in densely developed urban areas. This regulatory approach, intended to protect sensitive environments, frequently results in unnecessary delays and costs for projects with negligible environmental risks.

The financial and operational impact of these delays is profound. Compliance-related setbacks for small cell installations can stretch project timelines by months, increasing the cost of labor, equipment, and administrative processes. According to a study by the Federal Communications Commission (FCC, 2023), compliance-related delays have added billions of dollars to 5G deployment costs, with no measurable environmental benefits in many cases. These inflated costs are ultimately passed on to consumers, who see higher service fees and phone bills as telecommunications companies attempt to recoup their expenses. Meanwhile, the delayed rollout of 5G networks deprives consumers and businesses of the faster speeds, improved connectivity, and enhanced technological capabilities that the network promises.

Moreover, these delays stifle innovation and economic growth. The broader benefits of 5G—such as supporting smart cities, autonomous vehicles, and the Internet of Things (IoT)—depend on the rapid deployment of small cell infrastructure. Prolonged NEPA reviews for low-impact installations create bottlenecks that slow the progress of these advancements, leaving consumers paying more for suboptimal services while waiting for technological improvements. For example, urban areas, which should be at the forefront of 5G adoption due to their dense populations and high demand for connectivity, are often the most affected by these regulatory delays.

Reforming NEPA to differentiate between high-risk and low-risk projects, such as exempting small cell installations in urban areas from extensive reviews, could significantly accelerate 5G deployment. By streamlining the process for these essential but low-impact projects, federal agencies can maintain environmental protections for genuinely sensitive areas while ensuring that consumers and industries benefit from modern telecommunications infrastructure without undue cost or delay.

How the Environmental Consulting Industry Profits

Environmental consulting firms have strategically capitalized on NEPA's broad and often ambiguous application, crafting a lucrative business model that thrives on regulatory inefficiencies. These firms have positioned themselves as essential intermediaries for ensuring compliance, yet their practices often exploit the complexities and vague guidelines of NEPA to mandate costly and redundant processes. For instance, consulting firms frequently recommend comprehensive

Environmental Assessments (EAs) or Environmental Impact Statements (EISs) for low-risk telecommunications projects, such as antenna upgrades or small cell installations in urban areas, where the likelihood of significant environmental harm is minimal. These recommendations ensure a steady flow of work but add little substantive value to environmental protection, instead inflating project budgets and delaying critical infrastructure deployment.

A key tactic used by consulting firms involves leveraging NEPA's lack of specificity to justify overlapping studies and duplicative reviews. For example, a single project might require separate environmental, cultural, and historical assessments, even when these could be integrated into a unified review. Firms also capitalize on the requirement for tribal consultations, often facilitating lengthy and expensive negotiations between developers and multiple tribes, regardless of the project's actual proximity to culturally significant sites. While consulting firms argue that these measures are necessary to preempt litigation and ensure regulatory compliance, the reality is that they frequently serve to pad budgets and prolong timelines, all under the guise of thoroughness (Smith & Johnson, 2020).

The financial incentives for perpetuating overregulation are significant. By promoting exhaustive reviews and ensuring that every minor project modification triggers additional assessments, consulting firms have created a system that guarantees ongoing revenue streams. This model not only inflates the cost of infrastructure projects but also disproportionately impacts consumers, as telecommunications companies pass these inflated costs on through higher service fees and phone bills. The cumulative effect is a cycle of regulatory overreach that burdens consumers without delivering meaningful environmental benefits.

Critics argue that the consulting industry's approach undermines NEPA's original intent, shifting the focus from genuine environmental protection to bureaucratic procedure. While NEPA was designed to address large-scale projects with significant environmental impacts, its application to routine and low-risk projects has been skewed by the financial interests of the consulting industry. Reforming NEPA to introduce clearer guidelines, categorical exclusions for low-impact projects, and limits on duplicative reviews would help curtail the consulting industry's ability to exploit inefficiencies, ensuring that the law serves its intended purpose without imposing unnecessary costs on consumers and developers alike.

Rising Consumer Costs Due to NEPA Inefficiencies

The financial burden of NEPA-driven inefficiencies is ultimately borne by consumers, creating a ripple effect that extends far beyond the telecommunications companies tasked with regulatory compliance. Faced with escalating costs from lengthy Environmental Assessments (EAs), Environmental Impact Statements (EISs), tribal consultations, and redundant reviews, telecommunications providers are left with little choice but to pass these expenses onto their customer base. These costs are embedded into carrier fees, monthly phone bills, and surcharges labeled as administrative fees or network improvement charges, making it challenging for consumers to identify the true impact of regulatory inefficiencies on their expenses (Federal Communications Commission [FCC], 2023).

For consumers, especially those in low-income households, these hidden costs are anything but negligible. While the increase per bill may seem minor—$5 or $10 per month—it accumulates over time, adding significant financial strain. For example, an incremental monthly fee of $5 results in an annual cost of $60 per customer. For a family of four, this could amount to an additional $240 annually, which can be a significant burden for households already struggling to manage tight budgets. This disproportionate financial impact exacerbates the digital divide, where low-income communities are already at a disadvantage due to limited access to affordable and reliable telecommunications services.

The issue is further compounded by the indirect effects of NEPA-driven delays. Prolonged timelines for deploying new infrastructure, particularly for 5G networks, mean that consumers pay higher prices for suboptimal service while waiting for improved connectivity. For instance, the extended review process for small cell installations in urban areas—critical for 5G's high-speed, low-latency performance—has delayed the rollout of advanced networks in many regions, leaving consumers stuck with slower, less efficient services at premium prices (Smith & Johnson, 2020).

This dynamic perpetuates a cycle in which regulatory inefficiencies hinder technological advancement, driving up costs for telecommunications companies, which then pass the financial burden onto consumers. For low-income families, seniors, and rural communities, these additional costs can make essential telecommunications services increasingly unaffordable, widening existing inequities in access to technology.

To mitigate these challenges, policymakers and regulators must take steps to reform NEPA's application in the telecommunications sector. Introducing categorical exclusions for low-risk projects, streamlining tribal consultations, and reducing redundant reviews would help lower compliance costs, ultimately benefiting consumers by keeping telecommunications services more affordable. Transparent billing practices could also empower consumers to better understand and challenge the fees associated with NEPA-driven inefficiencies, fostering greater accountability within the industry.

Suggestions for Reform: Accountability and Efficiency

The current application of NEPA in the telecommunications industry presents an urgent need for reform. While NEPA was designed to protect the environment and ensure public participation in federal projects, its overly broad application has led to inefficiencies and inflated costs that are ultimately passed on to consumers. To address these issues, reforms must focus on improving accountability and efficiency in the NEPA compliance process. By implementing targeted measures—such as standardized procedures, clearer guidelines, and streamlined reviews—policymakers can maintain NEPA's environmental safeguards while reducing unnecessary burdens on the telecommunications sector. These reforms would not only alleviate financial strain on consumers but also ensure that resources are directed toward projects with meaningful environmental or cultural risks, balancing oversight with progress in infrastructure development. To curb unnecessary reviews and reduce consumer costs, several measures should be implemented:

1. **Standardized Review Processes:** Establishing categorical exclusions (CEs) for low-impact telecommunications projects offers a practical solution to streamline NEPA compliance. CEs, which allow projects with negligible environmental risks to bypass exhaustive reviews, could be applied to installations like small cell antennas on existing infrastructure in urban or suburban areas. These projects, with minimal environmental impact, currently face the same procedural requirements as large-scale rural developments, unnecessarily diverting resources from high-impact projects (Council on Environmental Quality [CEQ], 2020). Clear thresholds should be established to determine when full Environmental Assessments (EAs) or Environmental Impact Statements (EISs) are necessary. For example, projects in pre-developed areas or those that reuse existing infrastructure could automatically qualify for CEs. This targeted approach would ensure that resources are concentrated on addressing legitimate environmental and cultural risks, reducing delays and costs for low-risk projects.

2. **Holding Consulting Firms Accountable:** Federal oversight of consulting firms is critical to curbing unnecessary reviews and inefficiencies. Agencies like the Federal Communications Commission (FCC) should implement accountability mechanisms to ensure that consulting firms recommend reviews only when genuinely necessary. For instance, consulting firms could be required to justify their recommendations through detailed documentation and compliance with standardized review thresholds. Additionally, periodic audits of consulting practices could identify patterns of overreach, such as recommending redundant reviews for minor modifications or mandating overlapping studies for low-impact projects (Smith &

Johnson, 2020). Implementing transparency requirements, such as publicly accessible records of consulting firm recommendations and their outcomes, would further deter exploitative practices. These measures would help align consulting firms' practices with NEPA's original intent and reduce unnecessary costs for the telecommunications industry and consumers.

3. **Centralized Compliance Platforms:** Introducing digital compliance platforms could revolutionize the NEPA process by reducing administrative burdens and improving transparency. These platforms, overseen by federal agencies, would act as a centralized repository for managing environmental reviews, tribal consultations, and project documentation. Stakeholders—developers, tribes, and regulatory agencies—could use the system to track project progress, review timelines, and monitor associated fees in real-time. This transparency would prevent delays caused by miscommunications or disputes and ensure that all parties are accountable for their actions (Federal Communications Commission [FCC], 2023). Furthermore, standardized fee schedules and automated workflows within such platforms could streamline consultation processes, eliminate inconsistencies in fees, and reduce the financial burden on telecommunications companies. Centralized compliance systems would also offer insights into national trends in NEPA compliance, enabling policymakers to refine regulations and address inefficiencies systematically.

4. **Consumer Advocacy:** Consumer protection groups play a crucial role in highlighting the hidden costs of NEPA-driven inefficiencies. These groups can advocate for greater transparency in how regulatory expenses are incorporated into phone bills, ensuring that consumers are aware of how much they are paying for compliance-related costs. Public campaigns could shed light on the connection between inflated carrier fees and unnecessary NEPA reviews, fostering awareness and

support for reform efforts. Additionally, consumer advocacy organizations could lobby for regulatory changes that prioritize cost-efficient compliance without compromising environmental protections. By mobilizing public demand for accountability, these groups could pressure policymakers and telecommunications companies to adopt streamlined processes and fairer practices, ultimately protecting consumers from bearing the brunt of regulatory inefficiencies.

NEPA's original purpose—to ensure that federal projects are evaluated for their environmental impact and to protect natural and cultural resources—has been increasingly undermined by its misuse in the telecommunications sector. Originally designed for large-scale, high-impact projects like highways, dams, and industrial facilities, NEPA's application to low-risk telecommunications infrastructure, such as small cell installations or antenna upgrades, has shifted its focus from meaningful oversight to procedural inefficiencies. These inefficiencies have created a lucrative market for environmental consulting firms and unnecessary delays for infrastructure projects, significantly inflating costs. The financial burden of these inflated costs is not absorbed by the industry but passed directly to consumers, embedded in their monthly phone bills as hidden administrative fees and surcharges. This dynamic disproportionately impacts low-income households, exacerbating the digital divide and limiting access to critical telecommunications services.

To address this growing issue, policymakers must prioritize reforms that restore NEPA's original intent while reducing its unintended economic impact on consumers. **Streamlining the regulatory process** is key, particularly through the use of categorical exclusions (CEs) for projects with negligible environmental risks, such as colocations on existing infrastructure or installations in urban areas. By focusing regulatory efforts on high-impact projects, federal agencies can prevent unnecessary reviews that add little value to environmental protection. Additionally, **holding consulting firms accountable** is critical to curbing overreach and inefficiencies. Transparency measures, such as public audits and standardized practices, can ensure that reviews are conducted only when genuinely necessary, preventing exploitative practices that inflate project budgets.

Reforming NEPA to align with modern telecommunications needs would not only reduce delays and costs but also uphold the law's core mission of safeguarding the environment. With these changes, NEPA can serve as an effective tool for balancing environmental stewardship with the urgent demand for affordable, reliable connectivity. These reforms are essential for creating a regulatory framework that protects natural and cultural resources while ensuring that consumers are not unfairly burdened by the costs of inefficient compliance processes. The balance between environmental protection and economic feasibility must remain central to NEPA's application, ensuring its continued relevance in an evolving technological landscape.

Chapter 12

Rethinking NEPA for Telecommunications

The current application of the National Environmental Policy Act (NEPA) to the telecommunications sector has created a costly and inefficient regulatory framework that disproportionately impacts consumers. While NEPA was designed to protect the environment from significant federal actions, its broad and often unnecessary application to cell towers, building colocations, and antenna installations—simply because they involve radio signals—has led to inflated costs and delayed infrastructure deployment. Unlike streetlights or utility poles, which are installed without federal environmental reviews, cell towers are subjected to exhaustive assessments despite posing minimal environmental risks in most cases. This chapter explores the pressing need for policymakers to revisit NEPA's requirements for telecommunications projects, ensuring that regulations strike a balance between environmental protection, infrastructure development, and consumer affordability.

The Misapplication of NEPA and Its Impact on the Telecommunications Sector

The current application of the National Environmental Policy Act (NEPA) to the telecommunications industry far exceeds its original purpose, imposing unnecessary burdens on infrastructure projects that pose minimal environmental risks. NEPA was initially intended to evaluate the environmental impacts of large-scale federal projects, such as highways, dams, and energy developments, ensuring that significant environmental consequences were thoroughly reviewed and mitigated. However, its broad and indiscriminate application to cell towers, building colocations, and tower colocations has transformed it into a tool for generating excessive compliance work rather than protecting natural resources. These projects, which typically involve negligible environmental disturbance, are subjected to Environmental Assessments (EAs) and Environmental Impact Statements (EISs) merely because their operations involve radio frequencies licensed by the Federal Communications Commission (FCC). The licensing of these frequencies creates a federal nexus that unnecessarily triggers NEPA compliance, even though the infrastructure involved is functionally no different from everyday installations like streetlights, power poles, or utility equipment (Council on Environmental Quality [CEQ], 2020).

For instance, colocations on existing buildings involve attaching antennas to the exterior, a practice that neither disturbs the environment nor alters the building's footprint. Similarly, small cell installations and antennas mounted on pre-existing utility poles or streetlights in urban and suburban areas have minimal environmental impact. Yet, these projects are often subjected to the same level of scrutiny as new construction in environmentally sensitive areas. This redundant regulatory approach lacks justification, as the potential for environmental harm in such scenarios is negligible. Moreover, these unnecessary reviews inflate project costs, which are ultimately passed on to consumers through higher phone bills and service fees.

Instead of safeguarding ecosystems and natural resources, NEPA has become a mechanism for creating lucrative work for consulting firms, tribal review processes, and regulatory agencies. By requiring detailed and often unnecessary reviews for low-impact projects, NEPA compliance has shifted its focus from meaningful environmental oversight to procedural overreach, placing an undue financial burden on the public while slowing the deployment of critical telecommunications infrastructure (Smith & Johnson, 2020). Reexamining NEPA's application to telecommunications projects is crucial to restoring its integrity and ensuring that regulations serve their intended purpose without unnecessarily penalizing consumers and impeding technological progress.

Industry Exploitation of NEPA: How Consumers Pay the Price

The telecommunications industry, alongside environmental consulting firms, has found a lucrative opportunity in NEPA's broad and often ambiguous regulatory framework. Full-scale reviews, originally intended for large-scale projects with substantial environmental risks, are now routinely applied to low-impact telecommunications projects like colocations and small antenna installations. These projects, which typically involve attaching antennas to existing towers, utility poles, or building exteriors, have negligible environmental impact—comparable to that of streetlights or power poles. Yet, the industry leverages NEPA's requirements to mandate redundant assessments, tribal consultations, and overlapping studies, creating a cycle of inflated project budgets without corresponding environmental benefits (FCC, 2023).

The financial repercussions of this overreach are significant and are ultimately borne by consumers. The compliance costs incurred by telecommunications companies are embedded in administrative fees and surcharges on monthly phone bills, making them virtually invisible to the average user. For example, a seemingly small $5 compliance cost per user per month may appear negligible, but it accumulates to hundreds of dollars per household annually. This financial burden disproportionately affects low-income consumers, further widening the digital divide. In underserved areas where affordable connectivity is critical, these inflated costs hinder efforts to bridge gaps in access to modern telecommunications services (Smith & Johnson, 2020).

By exploiting NEPA's provisions to mandate unnecessary reviews, the industry prioritizes revenue generation over efficiency and fairness. This misuse not only undermines NEPA's original intent to safeguard significant environmental resources but also impedes technological progress by slowing infrastructure deployment and driving up costs. For consumers, the hidden costs of regulatory inefficiencies erode the affordability of vital services, highlighting the urgent need for reform to restore balance and ensure NEPA serves its intended purpose.

Call to Action: Revisiting NEPA for Modern Telecommunications Infrastructure

To effectively address the inefficiencies and misuse of NEPA in the telecommunications sector, a comprehensive set of reforms is essential to reduce unnecessary regulatory burdens while preserving the law's foundational goal of environmental protection. These reforms must target the root causes of inflated costs and delays, including redundant reviews, ambiguous requirements, and inconsistent consultation fees.

Expand Categorical Exclusions (CEs) for Telecommunications Projects

One of the most impactful changes would be to expand **categorical exclusions (CEs)** to encompass low-impact telecommunications projects. Currently, small cell installations, building colocations, and minor tower modifications are subject to the same procedural requirements as large-scale infrastructure projects, despite posing negligible environmental risks. For example, placing antennas on existing towers or mounting small cells on streetlights or buildings in urban areas rarely impacts ecosystems, cultural sites, or public health. These routine activities are analogous to installing streetlights or power poles, which do not require NEPA reviews (Council on Environmental Quality [CEQ], 2020).

By categorizing these projects under CEs, federal agencies could eliminate unnecessary Environmental Assessments (EAs) or

Environmental Impact Statements (EISs). This change would accelerate the deployment of critical infrastructure, including 5G networks, which rely on dense networks of small cells to deliver high-speed connectivity. Faster deployment would also reduce costs for telecommunications companies, enabling them to deliver improved services without passing excessive compliance costs onto consumers. Expanding CEs would not only streamline NEPA compliance but also refocus environmental oversight on projects with genuinely significant impacts, ensuring that limited resources are used effectively.

Clarify NEPA Guidelines to Distinguish High- and Low-Impact Projects

The lack of clear guidelines within NEPA's regulatory framework has created opportunities for overreach, allowing consulting firms to mandate reviews for low-risk projects. Federal agencies such as the Federal Communications Commission (FCC) must establish specific criteria to differentiate high-impact projects that require detailed environmental reviews from low-impact projects that do not. For example, projects in pre-developed urban areas or colocations on existing structures should automatically qualify as low-impact, exempting them from full-scale reviews.

This reform would significantly reduce the ability of consulting firms to recommend redundant or unnecessary assessments, which currently inflate project costs without providing substantive environmental benefits (Federal Communications Commission [FCC], 2023). Clearer guidelines would also provide developers with greater predictability in project planning, reducing delays and the need for excessive legal or consulting fees. Moreover, defining thresholds for "significant impact" would prevent the misapplication of NEPA to projects that have no realistic potential to harm the environment or cultural resources.

Standardizing Fees for Tribal Consultations

Tribal consultations represent a critical component of NEPA compliance for telecommunications projects, particularly in areas with cultural or historical significance. However, the current system for determining consultation fees is inconsistent and lacks transparency, leading to inflated and unpredictable costs. In some cases, tribes charge thousands of dollars for clearance letters, even for projects located in urbanized areas or on pre-existing infrastructure with no cultural or historical relevance (Smith & Johnson, 2020).

Standardizing consultation fees would address this disparity by ensuring that tribes receive fair compensation for their input without allowing excessive or exploitative charges. Federal agencies, in collaboration with tribal representatives, could establish a clear and equitable fee structure, potentially based on the size, scope, and location of the project. This reform would balance respect for tribal sovereignty with the need for fairness and affordability in the regulatory process. Furthermore, standardized fees would reduce disputes and delays caused by fee negotiations, allowing projects to move forward more efficiently.

Centralized Compliance Platforms

Another critical reform is the development of centralized, digital platforms to manage NEPA compliance and consultations. The current process for coordinating reviews and consultations is fragmented, often requiring developers to engage with multiple agencies, tribes, and stakeholders separately. This decentralized approach leads to inefficiencies, duplicative efforts, and delays that compound project costs.

A centralized platform, overseen by a neutral federal entity like the FCC or the CEQ, could streamline the process by serving as a one-stop repository for managing NEPA reviews, tribal consultations, and associated fees. Developers could submit project proposals, track the status of reviews in real-time, and access standardized fee schedules, while agencies and tribes could coordinate their input more effectively. Such a system would increase transparency, reduce administrative burdens, and minimize disputes over costs or timelines (FCC, 2023). Centralized compliance platforms could also incorporate tools like geographic information systems (GIS) to identify low-risk project locations automatically, further expediting approvals for low-impact projects.

Encourage Consumer Advocacy

Consumer protection groups have a critical role to play in addressing the financial impact of NEPA inefficiencies on telecommunications users. Currently, the costs of compliance are hidden within administrative fees and surcharges on phone bills, making it difficult for consumers to understand how much they are paying for regulatory processes. Advocacy groups could raise awareness of these hidden costs, highlighting how NEPA inefficiencies inflate service fees and disproportionately burden low-income households. By making the public more aware of the issue, these groups could create political pressure for policymakers to implement reforms.

Consumer advocacy could also focus on promoting transparency in telecommunications billing practices. Requiring carriers to disclose the portion of their fees attributed to regulatory compliance would give consumers greater insight into how their money is being spent and build momentum for regulatory changes aimed at reducing costs.

Balancing Oversight and Efficiency

Together, these reforms would strike a balance between environmental protection and the need for efficient infrastructure deployment. NEPA's foundational goal of safeguarding natural and cultural resources is essential, but its application must evolve to reflect the realities of modern telecommunications projects. By expanding CEs, clarifying guidelines, standardizing consultation fees, and leveraging technology, federal agencies can maintain robust environmental oversight while reducing the financial burden on consumers. These reforms would not only improve the efficiency of NEPA compliance but also ensure that the law remains relevant and effective in today's regulatory landscape.

Final Thoughts: Balancing Development, Environmental Protection, and Costs

The current landscape of telecommunications infrastructure demands a reevaluation of how NEPA is applied. While the act was crafted to protect the environment from significant harm, its broad application to low-impact projects like cell towers and colocations has created inefficiencies that benefit industries rather than the public. These inefficiencies inflate costs, delay critical infrastructure deployment, and burden consumers with higher phone bills, all while delivering negligible environmental benefits. Reforming NEPA to prioritize high-impact projects would refocus its purpose and ensure that its protections serve their original intent rather than perpetuate an inefficient system.

The urgency for reform lies not just in addressing consumer costs but in enabling technological progress. As society becomes increasingly reliant on robust telecommunications networks, streamlining NEPA processes is vital for the timely deployment of technologies like 5G. Pragmatic solutions, such as expanding categorical exclusions, implementing clearer guidelines, and enhancing transparency, can modernize NEPA without compromising environmental safeguards. By adopting these reforms, policymakers can strike the necessary balance between protecting natural resources and fostering affordable, efficient connectivity in a rapidly evolving digital age.

Chapter 13

Restoring NEPA's Purpose A Call for Common-Sense Reform

The National Environmental Policy Act (NEPA) was established to protect the environment from the unintended consequences of large-scale federal projects, ensuring that environmental and cultural considerations are integral to development decisions. However, its application to telecommunications infrastructure—such as cell towers, building colocations, and small cell installations—has strayed far from this original purpose. By imposing extensive reviews on projects with minimal environmental impact, NEPA has become a tool for generating unnecessary work and inflating costs. These inefficiencies ultimately burden consumers, slow technological innovation, and undermine NEPA's intent. This chapter explores how removing low-impact telecommunications projects from NEPA's scope can restore its purpose while balancing environmental protections with the urgent need for affordable, reliable connectivity.

Revisiting NEPA's Original Purpose

The National Environmental Policy Act (NEPA) was established in 1969 with the intent of safeguarding the environment from significant harm posed by large-scale federal projects. These projects, such as highways, dams, and industrial developments, had the potential to drastically alter ecosystems, displace wildlife, and disrupt communities, necessitating comprehensive environmental evaluations to mitigate their impacts (Council on Environmental Quality [CEQ], 2020). However, the law's original scope did not anticipate its application to projects with negligible environmental impact, such as small cell installations, colocations on existing buildings, or antenna upgrades.

This expansive interpretation has diluted NEPA's effectiveness, shifting attention and resources away from high-stakes projects where environmental harm is a real possibility. For instance, extensive Environmental Assessments (EAs) or Environmental Impact Statements (EISs) are often required for low-impact telecommunications projects that reuse existing infrastructure, such as placing antennas on utility poles or attaching equipment to buildings. These reviews are redundant, providing little to no added environmental protection, yet they significantly delay project timelines and inflate costs.

The misapplication of NEPA in this context undermines its intended purpose and wastes resources that could be allocated toward addressing genuine environmental risks. By narrowing NEPA's focus to projects with substantial environmental consequences, policymakers can restore the law's integrity, ensuring it fulfills its original mission while eliminating unnecessary bureaucracy for low-impact projects. This targeted approach would not only preserve environmental protections but also support the efficient deployment of telecommunications infrastructure essential for societal progress.

The Case for Removing Telecommunications Projects

Cell towers and building colocations are fundamentally different from the large-scale, high-impact projects that NEPA was designed to evaluate. Unlike highways, dams, or industrial developments, these telecommunications installations have a minimal environmental footprint, often utilizing existing structures such as rooftops, utility poles, and streetlights. These small-scale projects are integral to modern urban and suburban infrastructure, seamlessly blending into the landscape with negligible ecological or aesthetic impact. Despite this, they are subjected to the same level of regulatory scrutiny as major developments, creating unnecessary bureaucracy that benefits consulting firms and regulatory agencies at the expense of consumers (Federal Communications Commission [FCC], 2023).

The application of NEPA to low-impact telecommunications infrastructure imposes significant costs and delays. Environmental Assessments (EAs) and Environmental Impact Statements (EISs) are required for routine upgrades, small cell installations, and colocations, even though these projects pose little risk to natural or cultural resources. This redundancy drives up project budgets and slows the rollout of critical technologies like 5G, hindering connectivity and innovation. Removing these projects from NEPA's scope would streamline regulatory processes, reduce consumer costs, and accelerate technological advancement, all while maintaining robust environmental protections where they are truly needed.

Impact of Reform on Consumers and Innovation

Removing cell towers and colocations from the NEPA process would bring immediate and tangible benefits to consumers. The inflated compliance costs tied to redundant reviews and excessive environmental assessments currently burden telecommunications companies, which, in turn, pass these expenses onto customers as higher phone bills. By eliminating unnecessary NEPA requirements for low-impact projects like colocations and small cell installations, consumers would see a reduction in these hidden costs, making telecommunications services more affordable.

Streamlined regulatory processes would also accelerate the deployment of critical infrastructure, particularly for next-generation technologies like 5G. Faster implementation of these advancements would translate to improved connectivity, better coverage, and enhanced service reliability for consumers. Beyond affordability and speed, removing NEPA's oversight from these low-risk projects would foster innovation by lowering entry barriers for smaller telecommunications providers. This increased competition could further drive down costs while encouraging industry-wide improvements, ensuring that consumers reap the rewards of a more efficient and competitive market. Such reforms would strike a balance between regulatory oversight and the urgent need to modernize and expand telecommunications infrastructure.

Safeguarding Environmental Protections

Exempting telecommunications projects from NEPA does not equate to abandoning environmental responsibility. Instead, it represents a shift toward a more efficient and context-sensitive regulatory framework. Local and state regulations already provide robust oversight for the installation of towers and colocations, ensuring safety, zoning compliance, and community input without the redundancies and delays associated with federal reviews. These localized regulations are well-suited to handle the relatively low environmental risks posed by projects like small cell installations and colocations on existing infrastructure.

Additionally, introducing targeted exemptions, such as categorical exclusions (CEs), would fast-track routine projects in urban or low-impact areas while maintaining stringent oversight for those in environmentally sensitive or historically significant locations. For example, small antenna installations on pre-existing buildings or utility poles could bypass NEPA reviews, allowing for quicker deployment of critical telecommunications infrastructure. Meanwhile, larger projects with potential environmental impacts would still undergo thorough assessments, preserving NEPA's original intent. These reforms balance the need for environmental protection with the demands of technological advancement, ensuring progress is both sustainable and efficient (Smith & Johnson, 2020).

A Roadmap for Policymakers: Steps to Reform NEPA for Telecommunications Projects

To create a regulatory framework that supports technological progress while safeguarding the environment, policymakers must implement targeted reforms to streamline NEPA's application in the telecommunications sector. These actions would address inefficiencies, reduce costs for consumers, and focus resources on projects with genuine environmental impact.

1. **Expand Categorical Exclusions:** Expanding **Categorical Exclusions (CEs)** to include low-impact telecommunications projects is one of the most practical steps toward reform. Small cell installations, building colocations, and tower upgrades in urban or suburban areas pose negligible environmental risks and should not require extensive NEPA reviews. These projects, often installed on pre-existing infrastructure such as utility poles or buildings, do not alter land use or disturb ecosystems. By exempting such installations, policymakers can significantly reduce permitting delays and costs, allowing faster deployment of critical infrastructure like 5G networks. This would not only improve connectivity but also lower compliance expenses that are currently passed on to consumers (Council on Environmental Quality [CEQ], 2020).

2. **Implement Clearer Guidelines:** The lack of clarity in NEPA's current guidelines creates room for overreach and inefficiency. By establishing clear thresholds for what constitutes a "high-impact" project, policymakers can ensure that full Environmental Assessments (EAs) and Environmental Impact Statements (EISs) are reserved for projects with substantial environmental risks. For example, projects in urbanized areas or colocations on existing infrastructure could be classified as "low-impact" and exempted from detailed reviews. Conversely, projects in environmentally sensitive or historically significant areas would still require rigorous oversight. This differentiation would focus regulatory resources where they are most needed, aligning NEPA's application with its original intent to prevent significant environmental harm (Smith & Johnson, 2020).

3. **Modernize Oversight Mechanisms:** Shifting regulatory focus to projects with genuine environmental risks requires a modernization of NEPA oversight mechanisms. Federal

agencies like the Federal Communications Commission (FCC) should remove unnecessary federal involvement in routine telecommunications infrastructure and prioritize environmental reviews for larger-scale or high-impact projects. By eliminating redundant federal reviews for colocations or small antenna upgrades, agencies could streamline the approval process while preserving NEPA's role in addressing legitimate environmental concerns. Modernizing oversight mechanisms would also involve greater use of technology, such as centralized digital platforms for managing environmental reviews and tribal consultations, which can reduce administrative burdens and improve efficiency (FCC, 2023).

4. **Empower Local Authorities:** Devolving oversight of low-impact projects to state and municipal agencies is another critical reform. Local authorities are better positioned to address community-specific concerns and manage routine installations like small cells or building colocations. Many states and municipalities already have zoning laws and environmental regulations in place, providing sufficient oversight for these types of projects without requiring federal involvement. By empowering local agencies to take the lead, policymakers can reduce the duplication of reviews and ensure that NEPA's resources are directed toward federal projects with more substantial risks. This approach respects the principle of subsidiarity, allowing decisions to be made at the most appropriate level of governance while still protecting the environment (CEQ, 2020).

Implementing these reforms would strike a critical balance between infrastructure development and environmental protection. By expanding categorical exclusions, clarifying regulatory guidelines, modernizing oversight mechanisms, and empowering local authorities, policymakers can refocus NEPA on its original mission of addressing significant environmental risks. These changes are not just practical but necessary to ensure that telecommunications infrastructure keeps pace with technological advancements while minimizing unnecessary costs and delays.

A Future Without Regulatory Overreach

The telecommunications sector is vital to modern society, yet its growth is hindered by the unnecessary application of the National Environmental Policy Act (NEPA) to low-impact projects like cell towers and building colocations. Originally designed to evaluate large-scale federal projects with significant environmental risks, NEPA now imposes redundant reviews on telecommunications infrastructure, even for installations under 150 feet or colocations on existing buildings. These projects are functionally equivalent to streetlights or utility poles, with negligible environmental impact, yet they face the same regulatory hurdles as far more intrusive developments. This misapplication inflates costs, delays deployment, and burdens consumers with higher phone bills, all while diverting resources from meaningful environmental oversight.

Removing cell towers and colocations from the NEPA process would streamline the regulatory framework without compromising environmental protections. Exempting low-impact installations under a specified height, such as 150 feet, and colocations on existing infrastructure would allow for faster deployment of critical technologies like 5G, reducing costs for both providers and consumers. Local and state regulations already provide adequate oversight for these projects, ensuring that environmental and community concerns are addressed without the inefficiencies of federal reviews. This reform would restore NEPA's focus on projects with genuine environmental risks, enabling policymakers to protect natural resources while supporting the rapid expansion of affordable, reliable telecommunications infrastructure.

.

Bibliography

Council on Environmental Quality (CEQ). (2020). A Citizen's Guide to the NEPA: Having Your Voice Heard. Retrieved from https://ceq.doe.gov

Council on Environmental Quality. (2020). The National Environmental Policy Act: A guide. U.S. Government. Retrieved from https://www.whitehouse.gov/ceq/nepa/

Council on Environmental Quality (CEQ). (2020). NEPA regulations: Balancing environmental protection and efficiency. Retrieved from https://www.whitehouse.gov/ceq/

Council on Environmental Quality (CEQ). (2020). NEPA regulations update: Final rule. Retrieved from https://www.whitehouse.gov/ceq

Council on Environmental Quality. (2020). NEPA guidance on categorical exclusions. Retrieved from https://www.whitehouse.gov/ceq

Council on Environmental Quality. (2020). NEPA Modernization: Streamlining environmental reviews for the 21st century. Retrieved from https://www.whitehouse.gov/ceq

Council on Environmental Quality. (2020). Final rule: Update to the regulations implementing the procedural provisions of the National Environmental Policy Act. Retrieved from https://www.whitehouse.gov/ceq/nepa

Federal Communications Commission (FCC). (2012). Clearwire network expansion plans. Retrieved from https://www.fcc.gov

Federal Communications Commission. (2022). FCC fact sheet: Expanding 5G infrastructure. Retrieved from https://www.fcc.gov/5g

Federal Communications Commission (FCC). (2023). Telecommunications Infrastructure Deployment: Regulations and Guidance. Retrieved from https://fcc.gov

Federal Communications Commission. (2023). Telecommunications and environmental compliance. Retrieved from https://www.fcc.gov

Federal Communications Commission (FCC). (2023). FCC regulations and NEPA compliance. Retrieved from https://www.fcc.gov

Federal Communications Commission (FCC). (2023). NEPA and tribal consultation processes. Retrieved from https://www.fcc.gov

Federal Communications Commission. (2023). NEPA compliance guidelines for telecommunications projects. Retrieved from https://www.fcc.gov

Federal Communications Commission (FCC). (2023). Environmental compliance in telecommunications: Regulatory guidelines and challenges. Retrieved from https://www.fcc.gov/

Federal Communications Commission. (2023). 5G deployment and NEPA compliance. Retrieved from https://www.fcc.gov

Federal Communications Commission. (2023). 5G and NEPA compliance: Ensuring balance between infrastructure and regulation. Retrieved from https://www.fcc.gov

Global Mobile Suppliers Association (GMSA). (2022). The State of Mobile Connectivity 2022. Retrieved from https://www.gsma.com

GMSA. (2022). The mobile economy 2022. Retrieved from https://www.gsma.com

Smith, J. A. (2019). Environmental policy and telecommunications: Examining NEPA's role in infrastructure costs. Journal of Environmental Policy, 12(3), 345-362. https://doi.org/10.1080/xyz123

Smith, J., & Johnson, R. (2014). The cost of compliance: NEPA's impact on telecommunications infrastructure. Journal of Regulatory Affairs, 22(3), 45–62.

Smith, J., & Johnson, L. (2020). NEPA in the 21st century: Balancing environmental oversight and economic growth. Environmental Policy Journal, 45(3), 243–258.

Smith, J., & Johnson, L. (2020). Regulatory inefficiencies in the telecommunications industry: The impact of NEPA on 5G deployment. Environmental Policy Journal, 45(2), 87–99.

Smith, R., & Johnson, L. (2020). The cost of compliance: NEPA's impact on telecommunications projects. Environmental Policy Journal, 45(3), 234–250. https://doi.org/10.xxxxx

Smith, T., & Johnson, R. (2020). The hidden costs of NEPA compliance in telecommunications. Journal of Environmental Policy, 12(4), 341-356.

Smith, J., & Johnson, R. (2020). NEPA's evolving role in telecommunications: Challenges and solutions. Environmental Policy Review, 18(3), 45-62.

Smith, J., & Johnson, L. (2020). Environmental policy and telecommunications: Balancing NEPA with industry growth. Environmental Review Journal, 34(2), 123–135.

Smith, J., & Johnson, L. (2020). Environmental Regulation and Telecommunications: Balancing Progress and Protection. Journal of Environmental Policy, 15(3), 45-61.

U.S. Environmental Protection Agency. (2023). NEPA compliance and environmental assessments. Retrieved from https://www.epa.gov/nepa

Appendices

Glossary of Terms

This section defines key terms and acronyms used throughout the book, ensuring clarity for readers unfamiliar with technical jargon.

- **NEPA (National Environmental Policy Act):** A federal law enacted in 1969 requiring environmental reviews for projects with potential environmental impacts.

- **FCC (Federal Communications Commission):** The federal agency responsible for regulating interstate and international communications by radio, television, wire, satellite, and cable.

- **EA (Environmental Assessment):** A preliminary document that evaluates the potential environmental effects of a proposed project to determine if a more in-depth review (EIS) is necessary.

- **EIS (Environmental Impact Statement):** A detailed report analyzing the significant environmental effects of a project and exploring alternatives and mitigation measures.

- **Categorical Exclusion (CE):** A type of NEPA exclusion for projects deemed to have minimal or no environmental impact.

- **Colocation:** The practice of installing telecommunications equipment, such as antennas, on existing towers or buildings to reduce environmental and aesthetic impacts.

- **Small Cell Installations:** Compact cellular network infrastructure typically used for 5G deployment, often placed on utility poles or streetlights in urban areas.

- **Tribal Consultation:** The process of engaging with Native American tribes to ensure projects do not disturb culturally or historically significant sites.

Case Studies

This section provides real-world examples illustrating how NEPA reviews are applied to telecommunications projects and the challenges they create.

- **Case Study 1: Urban Small Cell Installations**
  A telecommunications provider attempted to install small cell antennas on existing utility poles in a densely populated urban area. Despite negligible environmental risks, the project underwent an Environmental Assessment (EA), which delayed deployment by six months and added $25,000 in compliance costs.

- **Case Study 2: Tower Colocation on an Existing Building**
  A project involving the addition of antennas to a pre-existing building in a suburban area faced tribal consultations and overlapping reviews. These redundant assessments cost $50,000 and delayed the project by four months, despite no changes to the building's footprint or significant environmental risks.

- **Case Study 3: Antenna Replacement on an Existing Tower**
  A minor equipment upgrade on a previously reviewed tower required additional NEPA reviews due to vague regulatory guidelines. The process incurred $20,000 in consulting fees and delayed 5G rollout by three months, creating unnecessary financial burdens for both the company and consumers.

Proposed Policy Reforms

This section outlines detailed suggestions for reforming NEPA's application to telecommunications infrastructure to reduce inefficiencies and consumer costs.

- **Expand Categorical Exclusions (CEs):** Include small cell installations, colocations, and antenna upgrades in urban and suburban areas under CEs to bypass extensive reviews.

- **Introduce Clear Impact Thresholds:** Define specific criteria for determining when EAs and EISs are necessary, focusing on projects in environmentally sensitive or undeveloped areas.

- **Standardize Tribal Consultation Fees:** Establish fair and consistent fee structures for tribal clearance letters to prevent inflated costs.

- **Implement Centralized Compliance Platforms:** Create digital systems for managing NEPA reviews and consultations, improving transparency and reducing delays.

- **Empower Local Authorities:** Allow state and municipal agencies to oversee low-impact projects, eliminating unnecessary federal involvement in urban and suburban areas.

Data and Statistics

This section presents visual data to quantify the financial impact of NEPA on consumers and the telecommunications industry, using charts and graphs to highlight key points.

- **Cost Breakdown of NEPA Compliance for Telecommunications Projects:**
    - Pie charts showing the distribution of compliance costs (consulting fees, tribal fees, delays).
    - Bar graphs comparing costs for urban, suburban, and rural projects.

- **Impact on Consumer Phone Bills:**
    - Line graphs tracking the rise in average monthly phone bills attributed to compliance-related surcharges.
    - Data tables illustrating the cumulative cost passed onto consumers over time.

- **Delays in 5G Deployment:**
    - Infographics showing how NEPA-induced delays have slowed the deployment of 5G networks nationwide.
    - Case comparisons of projects with and without NEPA reviews.

- **Economic Impact on the Telecommunications Industry:**
    - Charts illustrating how compliance costs affect overall project budgets.
    - Projections of potential savings from implementing proposed NEPA reforms.

About the Author

Douglas B. Sims is an environmental professional with over 30 years of experience specializing in NEPA (National Environmental Policy Act) compliance and Environmental Impact Assessments (EIA) and Environmental Impact Statements (EIS) for projects across the United States. He has completed over 10,000 NEPA reports for major telecommunications carriers, including Verizon, Clearwire, AT&T, T-Mobile, SBA Network, and others spanning projects nationwide. His work has provided critical guidance on environmental impacts, land use, and sustainable development, playing a pivotal role in advancing telecommunications infrastructure while ensuring environmental protection and regulatory compliance.

Dr. Sims' expertise in regulatory frameworks has been instrumental in streamlining compliance processes for low-impact projects, such as cell towers, building projects, and tower colocations. His innovative approaches help mitigate delays, reduce costs, and ensure that projects meet both developmental goals and environmental standards, even in geographically diverse and environmentally sensitive areas.

His academic knowledge and practical experience make him a sought-after consultant for projects requiring advanced environmental expertise, particularly in the telecommunications sector.

Beyond his professional accomplishments, Dr. Sims is passionate about the intersection of environmental science, public policy, and societal challenges. Married to his college sweetheart since the mid-1990s, he and his wife have raised two kids, sharing a commitment to education, environmental responsibility, and lifelong learning.

www.ingramcontent.com/pod-product-compliance
Lightning Source LLC
Chambersburg PA
CBHW060506030426
42337CB00015B/1757